Technology Systems for Small Farmers

Published in cooperation with IFAD,
the International Fund for Agricultural Development
Rome, Italy

Technology Systems for Small Farmers

Issues and Options

EDITED BY
Abbas M. Kesseba

WITH A FOREWORD BY
Idriss Jazairy

LONDON AND NEW YORK

First published 1989 by Westview Press

Published 2019 by Routledge
52 Vanderbilt Avenue, New York, NY 10017
2 Park Square, Milton Park, Abingdon, Oxon OX14 4RN

Routledge is an imprint of the Taylor & Francis Group, an informa business

Library of Congress Cataloging-in-Publication Data
Technology systems for small farmers : issues and options / edited by
Abbas M. Kesseba ; [foreword by Idriss Jazairy].
p. cm.
ISBN 0-8133-7925-3 (pbk.)
1. Agriculture—Appropriate technology. 2. Agricultural extension
work. 3. Agricultural innovations. 4. Agriculture—Technology
transfer. 5. Agriculture—Research. 6. Farms, Small. I. Kesseba,
Abbas M.
S494.5.A65T43 1989
338.1'6—dc20 89-27248
CIP

ISBN 13: 978-0-367-28971-3 (hbk)
ISBN 13: 978-0-367-30517-8 (pbk)

CONTENTS

TABLES

FIGURES

FOREWORD

Creating a Link Between Small Farmers and Agricultural Research

Far too often well-intended development projects fail because the technology being applied does not fit the needs or capabilities of the local farmers. Many developing countries are littered with broken-down machinery which after a few months or years of operation sit idle for lack of maintenance. Far too many seemingly successful development projects collapse once the initial funds dry up because the project implementation staff were unable to secure the participation of the intended beneficiaries. This unfortunate reality has led the International Fund for Agricultural Development (IFAD) to focus increased efforts on identifying the root causes of the problem and, once identified, on implementing cost-effective alternative methods which increase the possibility of achieving the goal of sustainable development.

Traditionally, agricultural research and technology is delivered to farmers by extension workers based on what the market has to offer and how the market can adapt to the needs of the farmers. In the industrialized world, this system has worked well. Farmers have gradually incorporated technological advances, such as increased mechanization, irrigation and the use of hybrids, fertilizers, pesticides and herbicides into their farming practices, dramatically increasing productivity. But the technology developed in a rapidly advancing industrialized society often has limited application in different climatic zones or remote areas of Africa, Asia and Latin America. Moreover, the relatively homogeneous socio-economic structure of the industrialized countries is unparalleled in developing countries marked more by social, economic and political diversity than similarity.

As a result, extension services based fundamentally on the delivery of research and technology, conceived and developed within the agricultural context of the industrialized countries, cannot always be applied successfully in the developing world. This is especially true when extension services are required for remote, unschooled, impoverished, and often subdued rural smallholder or landless farmers in marginal lands subject to drought, floods, disease and insect invasions. Thus the challenge in both research and extension is to find out what the people need and determine how it can be delivered best.

With these considerations in mind, IFAD's investment in agricultural research has moved away from a top-down mechanistic and task-oriented exercise towards a demand-driven problem-solving approach directly related to the needs of the rural poor and their environment.

To begin with, this approach mandates that the research structure be close to the beneficiaries and be low in cost, with minimum reliance on external expertise, capital and equipment which would result in overdependence. Most importantly, however, the research structure must be based on the small farmers themselves. Their input is required to define the problems as well as to select, apply and evaluate alternative solutions. In this way, farmers and researchers can work together as equal partners, building an internally sustainable system which ensures a steady flow of information from the laboratory to the land and from the land to the laboratory.

The extension worker plays a critical role in the process by encouraging the continued participation of the small farmers and in ensuring that the applicable research is delivered to the farmers in a usable and acceptable manner. For example, in certain countries, social and cultural practices sometimes prevent male extension officers from working effectively with women farmers. In the Eastern Regional Agricultural Development Project in Yemen, P.D.R., IFAD has overcome this problem by including a significant number of women extension workers in the project who are able, for the first time, to provide women farmers with information vital to enhancing agricultural production. These same extension workers provide training in basic health, child care and literacy. Since, in many IFAD projects, literacy is very low, extension workers are also encouraging learning-by-doing methods which ensure participation.

Another recent example of IFAD's demand-driven approach is the Agricultural Development Programme in the Highlands of Madagascar. The programme commences with the extension agents identifying with the local farmers their needs and concerns. These are then communicated to the research specialists who develop techniques which are in turn delivered to the farmers by the extension agents. This approach, placing the smallholder at the centre of the process, is proving to be a promising alternative to traditional methods -- particularly for the resource-poor rural farmers in marginal areas.

A key challenge in focusing on the demands of the client presents itself, however, with the realization that the needs of each beneficiary group are different, requiring creative and innovative solutions in research and extension. For example, in an IFAD-sponsored project in the remote Peruvian highlands, after a careful evaluation of the needs of the peasant communities and a detailed cost-benefit analysis of the social as well as economic implications of the project, it was agreed that a live-in extension worker would be assigned to each village. The villages are participating in a cost-sharing arrangement for the extension services with the goal of achieving in time a peasant controlled and financed extension programme.

Emphasis on adapting research and extension services to meet the needs of the rural poor is becoming increasingly relevant every day, as burgeoning populations create additional stress on the world's forests and marginal agricultural lands. New extension techniques are needed, which build on the local people's traditional conservation practices and provide acceptable additional methods for the protection and enhancement of their natural resource base.

In recent years, other international, regional, national and non-governmental organizations have also been developing new approaches to fit research and extension services to the needs of smallholder farmers. The findings and lessons learned need to be consolidated and disseminated so that all can share in the acquired knowledge. This volume is an important first step in collecting and analyzing these experiences. It represents, in my view, a valuable contribution to furthering our understanding of the complex challenges involved in meeting the needs of the rural poor in their struggle to achieve sustainable development.

Idriss Jazairy
President
International Fund for
Agricultural Development

INTRODUCTION

The aim of this volume is to seek a better understanding of the issues and options involved in the generation and transfer of technology to poor small farmers. This book evolved partially from the proceedings of a seminar organized by IFAD on the Generation and Transfer of Technology for Poor Small Farmers held in Korea in June 1988. The following chapters represent a wide range of views and experiences from authors who, through their work with bilateral donor agencies, multilateral financing institutions, international institutes and UN agencies, have accumulated a wealth of knowledge on the subject. While these authors share common concerns about the generation and transfer of sound technology to enhance small farmer welfare, they represent different schools of thought as to possible approaches to resolving the central issues associated with the subject. It is hoped that the diverse, though profound, views presented here will help to promote and improve understanding of this complex and difficult area of Development.

This book is intended to provide a fresh opportunity to develop guidelines for the future design and implementation of rural development investment projects in which research and extension will interact with the small farmer in such a way that the latter serves as the lynch-pin in the system. As such, it represents one of the first attempts to view the various issues related to agricultural research and extension, solely from the perspective of the small farmer.

THE SETTING

Technology transfer systems date back to time immemorial. Even as man was beginning to practise sedentary agriculture, traditional systems of extension were evolving, shaped mainly by the prevalent socio- economic conditions and cultural traditions. Until recent times, these systems have kept pace, through the ages, with the demand for increased production and the need to enhance resource productivity. At the same time, these traditional extension systems, mainly in the developing regions, have yielded to newer forms which place emphasis on the expansion of cash crops and exportable agricultural commodities and cater to the markets of industrialised regions.

However, as agriculture developed in these (latter) regions as well, the pattern of the evolution of extension systems was very different and as such this was responsible for the emergence of a producer-based demand-driven extension system. This forms the basis of the present-day mode of agricultural extension in most developed countries and has been widely regarded as the most efficient form of extension.

In most of the developing world, however, extension systems still evolve around top-down supply-driven extension approaches which do not adequately address the socio-economic situation of resource-poor smallholder producers or their requirements. The major challenge in this context is, therefore, to develop appropriate technology transfer systems in developing agricultural economies. This in turn, entails the generation of a technology/extension process that takes into account the smallholders' perceptions and requirements. The issue is to ensure the effective participation of small farmers in the system, through a two-way interaction/feedback with researchers and extension agents.

The smallholder farming system is highly complex. It forms part of a sector characterized by a fragile, heterogeneous agro-sociological environment. Since the interaction between land resources and family labour is often the entire basis of operation, the small farmer generally tends to resort to the pursuit of a risk-averse production strategy, in an effort to eke out a survival.

Scarcity of available resources and vulnerability to exogenous factors, such as climate, market fluctuations, fiscal policies, and opportunity cost of labour, among others, make the production system highly unstable.

In spite of the above constraints, smallholder farmers represent a major, unexploited potential for food production and economic growth in many developing countries. The traditional production systems of small farmers are basically rational. However, the impact of externalities and demographic pressure often results in the breakdown of those traditional systems. This undermines farmers' potential and limits their options to pursue their own production strategies and decision-making.

Nevertheless, small farmers are clever, discerning economists, who are often enthusiastic to participate in improved farming practices and adopt innovative technical packages, if the economic advantages of these are clearly demonstrated.

This outlook, backed by the small farmers' production potential capacity, should be the major driving force behind technology generation and transfer systems. This is crucial for ensuring the increased agricultural productivity of small farmers and unleashing their vast potential to contribute to the overall economy and the popular national goal of achieving growth, with equity, in many developing countries.

The general failure of developmental efforts to adequately address the central concerns of the small farmer, is reflected by major drawbacks experienced during the design and implementation of research and extension projects. These emerge from supply-oriented strategies frequently followed by donors and national policy makers, which result in the inadequate response of research and extension systems to the specific needs of the target groups in the small farm sub-sector.

The demand for new technology and its adoption differs greatly within the universe of small-farmer production systems. Research needs to address this diversity, while the technological innovations offered must be consistent with the prevailing farming system. However, technological innovations are seldom aimed at resource-poor farmers. They are not cost effective in terms of low requirements for expensive external inputs, nor do they incorporate appropriate mixes of farm-grown and market-supplied inputs, which the resource-poor farmer can afford. These innovations should offer the possibility of increasing productivity without compromising the sustainability of the system. Consequently, technology systems which merely depend on the use of costly inputs have an inherent bias in favour of large farmers and commercial crops. As a result, newly generated technology may not always be relevant to the needs of poor farmers.

Technology systems also rely on constant intensification of resource use without giving due consideration to the conservation of the resource base. This is a critical issue in the development of marginal lands. In the absence of a conservation-based technology, and given the linkage between poverty and environmental degradation in marginal areas, further degradation leads to the abandonment of agricultural land, the rise of rural out-migration, and the subsequent creation of urban slums.

Effective small-farmer participation in the technology development and transfer system has a positive influence on the system's priorities and programmes. Mobilizing farmers' organizations in support of research and technology transfer systems is often found to be essential because it facilitates the implementation of decentralized research and can speed up information flow. Moreover, it may eventually become, directly or indirectly, a lobbying power which could ensure a better share of resources for these sub-sectors through the redirection of rural services which respond to small farmer needs in both high potential as well as marginal areas.

Unfortunately, such participatory structures and pressure groups are virtually non existent in many developing countries, especially in the context of disadvantaged peoples, whose requirements are often overlooked by national research and extension systems. One of the major factors responsible for this is the fact that the research and extension needs of disadvantaged groups -- resource-poor farmers, women, indigenous tribal populations, etc. -- can be of a highly specialized nature and not necessarily in conformity with the overall national needs of research and extension.

It should also be recognised that bringing research and technology to the small farm, however appropriate from the strategic point of view, is highly resource intensive, both financially and in terms of management requirements. Moreover, government commitment to apportion increased funds for agricultural research and extension is often not forthcoming, particularly under the restrictions imposed by structural adjustment programmes. While national development plans generally emphasize goals for achieving self-sufficiency in food production, they often fail to define or recognize the important role that research and extension can play in achieving that goal. Therefore, national research and extension (R&E) is often not endowed with the necessary resources and are frequently found lacking in institutional support to achieve those objectives. In many instances there is no specific policy framework for an effective technology transfer system, which could assist in setting research and extension goals, identifying the organizational and institutional arrangements to achieve those goals and creating better linkages between research, extension and the farmers themselves.

Technology generation and transfer systems as part of rural development projects do not always attract adequate government commitment to sustain these investments. Furthermore, investment projects which are either prepared by governments or supported by external donors, rarely define or articulate the goals to be pursued by the research and extension system they propose to support.

This results in a lack of clarity in objectives and is a further cause of the inadequate impact of the operating system on smallholder development.

AGRICULTURAL RESEARCH

Throughout the next decade, agricultural research will continue to face important challenges in the developing countries, especially in developing sustainable technologies for the heterogeneous agro-ecological and socio-economic conditions prevailing in those countries. Chapter one outlines a research agenda to address some of the central issues constituting the technological challenges discussed therein. In summarising IFAD's role in the generation of sustainable agricultural technologies, it sets the tone of the book, which is to promote the appropriate technological change within the context of small resource-poor farmers.

The importance of the small farm sector is further elaborated in Chapter two. Both chapters provide the rationale for a systems approach to agricultural research and suggest that in order for such research to become more relevant to the problems of resource-poor farmers, deliberate effort is needed, particularly through adaptive, on-farm research, to identify the problems of small farmers and include them in research design. By actively participating in collaborative feedback systems, researchers have, in many instances, been able to develop innovations appropriate to the needs of poor farmers. Furthermore, effective means of collaboration and joint planning need to be developed among research institutions, universities, the private sector and farmers' organizations.

This is also useful in the context of frontier technologies which have a great potential to improve traditional crops grown by resource-poor farmers. The private sector plays an important role in this area, for example, by developing methodologies for Biotechnology research for the genetic improvement of crops and livestock. Joint ventures between International Agricultural Research Centres (IARCs) and the private sector have the potential to apply such research to commodities that are produced and consumed by the resource-poor farmers.

Broad-scale, full-time, interdisciplinary research collaboration is difficult to mount and sustain. Furthermore, the lack of adequate linkage between on-farm research and extension is a chronic problem area in developing countries. Chapter three introduces an on-farm client-oriented research (OFCOR) approach which, when implemented within the context of several national agricultural systems (NARS), has successfully demonstrated that it can contribute to the transfer of technology by systematically diagnosing the constraints faced by farmers and by testing alternatives in small farm conditions. It has thus ensured that the farmers who had helped to conduct the research in the first place, could also benefit from its fruits.

The most important single factor which OFCOR introduces into a research system is farmer participation which keeps the programmes focussed on farmer needs. Innovative mechanisms need to be evolved to ensure proper coordination of field research at various levels and on a continuous basis. However, OFCOR cannot function effectively as an isolated research programme. It needs to be integrated within research systems, especially in areas where an appropriate extension message or technical package has yet to be developed.

An effective OFCOR system would necessarily involve researchers from a wide range of disciplines to undertake research in farmers' fields. Social scientists such as rural sociologists, for example, should play a greater role in research programmes to ensure greater sensitivity to the problems of the small farmers in different socio-cultural settings.

In whatever way it is organised, there are many crucial research issues which need to be addressed in the context of the small farmer. Gains from the adoption of technology may achieve higher levels of agricultural production but may still be inadequate to meet the levels of per capita self-sufficiency or provide household food security. New production technologies may, therefore, need to focus not merely on increasing labour productivity per se but also on the generation of economically meaningful off/on-farm employment. This would increase the effective demand for food and the purchasing power of small farmers, who would otherwise have been merely producing for subsistence.

Although new technological packages need to have an inherent element of resource management, (as discussed in Chapter one) an even more difficult challenge to research is that of exploring technologies which can not only restore but also enhance the productive capacity of the resource bases on a sustainable basis in order to cater to the demands exerted by growing demographic pressures.

Research has not yet focussed adequately on traditional technologies and on the production of basic food crops which are both relevant to the household needs of small farmers and conform to their dietary patterns. Production technologies have had a strong market orientation and have generally been biased against household consumption. There has been limited focus on post-harvest and processing technologies, and consequently a general lack of research focus on crucial issues such as seasonal food insecurity and nutritional requirements at the micro level. As a result relevant technology utilisation has been undermined.

AGRICULTURAL EXTENSION

The main purposes of agricultural extension are those of serving as an intermediary between the research system and farmers, delivering tested technology to farmers, and providing a feedback of farmers' problems to research so that it can be made more relevant to farmers' needs. Chapter four provides an overview of agricultural extension systems as well as a critical review of recent developments relating to extension. Emphasis has frequently been placed on the need to expand the set of functions which agricultural extension can perform, from mere information delivery to include farmer education, problem solving, adaptive research (and feedback) as well as marketing. In this way, the extension approach would not be confined to production aspects alone but would also focus on the total performance of the sector.

Poor impact from extension may sometimes result, not so much from any defect in the system as from a failure to ensure effective programme implementation. Extension management and organization are critical areas which need to be strengthened at all levels. Reforms on this front can be sustained only if agricultural extension is wholly integrated with the mainstream of the delivery system. It should be dynamic and flexible as well as sustainable over time. However, a broad-based, comprehensive extension approach is very uncommon.

The approach followed by most extension systems, in the past, lacked sufficient sensitivity to farmers' demands and failed to adapt to the socio-economic conditions of different groups of farmers. There has been limited use of various extension methods, (such as those associated with the use of contact farmers) and insufficient links with farmers' groups, the media and communication support systems. Besides State involvement, a variety of private sector agencies, such as input manufacturers, commodity boards, cooperatives and NGOs, have a potentially powerful role to play in extension and technology transfer activities. This potential, however, remains largely unexploited.

It is impossible to have a single, generalized extension system/approach suitable to all farming communities since every country has unique manpower and institutional resources, not to mention differences in agro-ecological conditions. However, extension systems have not been viewed in the context of the location- specific situation and this has frequently resulted in inappropriate organizational and programme focus on small farmer problems. Furthermore, the extension systems have usually been considered as ends in themselves and not as a means to an end. Mechanical application of extension systems by transferring and replicating models from one country (India) to another, has led to overlooking the essential elements of extension which are often required in different situations. This is revealed in Chapter Five which describes how the extension system has evolved in a country which has the largest extension system in the world today.

An examination and evaluation of past extension management experience reveals that no single pattern for extension can satisfactorily cope with the various agro-ecological conditions, socio-economic environments and administrative structures encountered in serving poor small farmers. In describing the World Bank's approach to agricultural extension, mainly the widely adopted Training and Visit (T&V) system, Chapter Six discusses the fundamental principles on which this system is based. It also discusses the orientation toward farmers the T & V system seeks to adopt, although at times it often fails to achieve it due not in the least to a general lack of compatibility between the T & V principles and certain aspects of their management and implementation.

However, it is recognized that the T & V system is only one of several ways of organizing human resources to carry out the extension process. Taking into account factors such as the variety of client characteristics and the constraints they face, as well as the availability of appropriate/adaptable technology, it has become necessary to modify the approach towards implementing the T&V model. This need has resulted not so much from any defect in the T&V system itself as from its failure to ensure effective participatory development, due mainly to the sophisticated management system it entails. Such a system is not realistic in the context of the poorly motivated extension staff, who are usually entrusted with its operation in most developing countries. The conventional T&V model would be more useful if it showed greater participatory emphasis i.e., if it were less directive (top-down, supervisory and supply-driven) and more participatory (bottom-up and farmer-demand driven).

The essential elements of the T&V system are not compromised by thus enlarging its horizon. Such restructuring retains the important characteristics of the system, one of which is training, which is crucial for the success of research and extension. In this context, informal training is just as important as formal methods of instruction, particularly where small farmers are concerned. However, the need for training to be vocational in nature, with a focus on skills development, has seldom been emphasised. The needs of small farmers - which should be the prime consideration in determining the scope, content and location of the training programme - have been neglected and this is a crucial issue which still needs to be resolved.

The role of agricultural extension, therefore, goes beyond technology transfer to include human resource development of the rural community. However, this is seldom recognized by policy makers since the economic and social pay-off from investments in extension services is not readily perceived and is therefore difficult to evaluate objectively. In discussing agricultural extension investment, Chapter Seven clearly points to the need for redressing this issue within the context of a national agricultural development policy framework. The chapter also calls for a better recognition of the contribution research and extension investments (especially investment for extension infrastructure) make in increasing agricultural productivity and output.

In general, extension systems have not been subjected to periodic assessment and modification with the result that the basic need for a professional extension service is often not properly assessed. Consequently, high quality, relevant technology development on the part of research, adequate programme and operating resources and systematic farmer contact are compromised. Lack of assessment has also led to the design of faulty extension and delivery systems and the failure of development projects to obtain their intended impact. It also emphasizes the importance of monitoring and evaluation within extension in order to provide feedback to extension management.

The main challenge in the context of extension systems is manifest in the formidable task of designing comprehensive situation-specific systems which take into consideration the above issues and are based on the desirable elements of participatory development. The efficient expansion in the generation and utilization of new agricultural production technology is contingent on the institutional development of research and extension systems, the need for which is discussed in Chapter Eight. Finally, the significance of economic policies and principles in making relevant new technology profitable in the context of small farmers, is discussed in Chapter Nine.

The foregoing introduction reflects the complex nature of Research and Extension issues which have not yet received adequate attention from the international development community. We intend to elaborate on these issues in the following chapters and draw the attention of research institutions and development agencies to the central issues involved in the generation of Technology Systems so that they can provide appropriate technologies and ensure their efficient transfer to poor small farmers.

Abbas M. Kesseba
Rome, August 1989

CHAPTER ONE

IFAD'S ROLE IN THE GENERATION OF SUSTAINABLE AGRICULTURAL TECHNOLOGIES

Abbas M. Kesseba and Shantanu Mathur[1/]

BACKGROUND

Despite significant gains in foodcrop productivity in recent years, the food situation in most developing countries remains critical. It has been estimated by the Food and Agriculture Organization of the United Nations and the World Bank that about 450 million people are on the verge of starvation and another one billion are undernourished. The global food situation is likely to become even more precarious in the closing decade of the century. Total food demand is expected to increase rapidly, particularly in the developing countries of Asia, Africa, Latin America and the Caribbean, due to rising nominal incomes and urbanization. Moreover, the situation is expected to be further confounded as a result of demographic pressure in the developing countries.

The shortfall in food availability might be compensated for, in part, through the expansion of the area under production, wherever this is still possible. However, this would become increasingly difficult to

1/ Coordinator and Research Associate, respectively, Technical Advisory Unit, International Fund for Agricultural Development, Rome.

achieve given the scarce land reserves which limit the scope for expansion. Thus, the major gains would have to come from sustained increases in crop and livestock productivity. The potential for increased crop productivity lies primarily in the small farmers' sub- sector because, so far, they represent an untapped resource for increased agricultural and food production. This entails a formidable technological challenge and indicates that substantial support is needed for agricultural research in order to generate innovative, appropriate technologies which can be adapted to small farmers' resource-base and socio-economic conditions.

The International Fund for Agricultural Development (IFAD), as the UN financial institution with a major focus on food and agricultural development by small farmers, would inevitably have to play a leading role in promoting technological change capable of meeting the food challenge and strengthening the attack on rural poverty.

This paper explores some of the central issues constituting the technological challenge for research and briefly describes the Fund's increasing role in supporting certain aspects of agricultural research. It also examines some critical issues and options which should feature prominently in its future agenda for further financial support to agricultural research.

IFAD'S MANDATE AND LENDING CRITERIA

IFAD's Lending Policies and Criteria recognizes the close linkages between hunger and rural poverty, as reflected in its mandate for increasing food production, alleviating rural poverty and improving nutrition for small farmers and poor rural households.

The Fund seeks to foster development at the level of the rural household, largely through self-help. Its prime focus is on the rural poor and the constraints they face. In rural areas, however, self-help is often limited by a lack of access to appropriate technological innovations capable of increasing returns from family labour, raising the level of its collective agricultural productivity, and thus progressively freeing it from the bonds of poverty in which it would otherwise become increasingly trapped.

IFAD's development strategies and the investment projects it finances are consequently dependent on the generation of technological innovations which can be adopted by poor rural households, especially in the least advantaged areas. However, the Fund recognizes that its strategies cannot fully succeed unless mechanisms exist for the generation and delivery of these <u>appropriate</u> technological innovations through research and extension.

Experience so far indicates that IFAD's support for agricultural research has had a catalytic effect in directing additional resources and manpower towards the problems of the rural poor. Where technical innovation has been developed and its viability proved, it has been linked to the formulation of investment projects for wider diffusion among its target groups with a minimum time-lag.

The Fund's explicit concern for research directed at the welfare of lower income groups in rural areas of developing countries, is also reflected in its "Lending Policies and Criteria" which states that "the food problem of the poor may be approached from different angles: e.g., by encouraging research and extension specific to the production of foods consumed by the poor; by taking development programmes to where poor people live; by researching, developing and extending technologies which increase employment while raising the productivity of capital and lands; and by favouring policies which guarantee equitable income to food growers and associate them with the overall benefits of development". The Fund has accordingly been supporting research programmes which adhere to both efficiency and equity goals through selected activities in international, regional and national research institutions. Furthermore, through its financing, the Fund has emphasized the need to focus on research which leads to new production parameters in favour of the small farmers and the landless in the developing world.

In targetting its development strategies to meet the needs of the small farmer, IFAD has supported adaptive research within the context of the farming system through which many technological innovations have emerged. However, the Fund realizes that if those innovations are to take full advantage of the advancing frontiers of science and technology, it must lend its full support to all aspects of the research and extension on which the future well-being of poor farmers depends. The Fund also recognizes that to support merely the final stages of applied and adaptive research on a short-term basis will not suffice. Long-term support for the whole research spectrum is essential to ensure both continuity of effort and the flow of new knowledge into the development process. Effectively speaking, development without closely associated research is severely constrained from the outset and is likely to lack vigour and dynamism. The Fund is concerned that such development, when not guided by an adequate technology system, will not achieve its desired impact and instead create a dependency on external sources for its technological input, which may not always be relevant to its target groups - the small farmer.

Thus, IFAD's stated goals for increased food production, alleviation of poverty and improvement of the nutrition level of poor rural households, depend on the availability of sustainable technologies which can only be developed through a continuous research effort carefully targetted to the issues facing specific poverty groups. The viability of the IFAD's investments is therefore crucially dependent on the extent of its support to agricultural research which has to meet the critical demands of the Fund's development efforts.

THE TECHNOLOGY CHALLENGE

Although the past fifty years have witnessed enormous advances in world agriculture, both hunger and poverty continue to threaten the well-being of millions of people in developing countries today. Moreover, the present rate at which the earth's surface is being denuded of its natural vegetation, cannot continue for much longer without irreversibly destroying the quality of the environment and the resource base on which its productivity depends.

As arable land reserves become increasingly scarce, the scope of expansion of cultivable areas which can increase agricultural production will be vastly reduced. The seemingly logical way out (and perhaps the easiest solution) has been to intensify agricultural production. Thus, within little more than a century, increasing agricultural production will have been transformed from a process of expanding the area under cultivation to one of intensifying production in areas which will be increasingly under pressure from a growing population and the need to preserve the quality of the environment.

To a large extent, problems associated with the intensification process are already alarmingly pronounced in many parts of the developing world. Given the present state of technology in high potential and fertile areas, the threshold of productivity levels has been reached, and further intensification (caused by demographic pressure or other factors) would entail diminishing returns to scale of production. In the fragile ecosystems, the current production levels (on which the substantial population in marginal lands subsist) cannot be sustained over time because productivity levels are rapidly on the decline. In the former case, production systems are manifested by newer, capital-intensive technologies which indulge in mining the land; while, on the marginal lands, the intensification of resource use (and its consequent degradation) is survival induced. In both cases, the major concern is the stress exerted on the resource base and the accompanying environmental degradation caused by these production systems.

Both types of production systems described above exemplify unsustainable agricultural development and emphasise the need for developing suitable technologies which can enhance productivity under diverse enviromental conditions, while, at the same time, incorporating considerations relating to conservation of the resource base. This represents the major technological challenge of the coming decades.

SUSTAINABILITY OF AGRICULTURAL PRODUCTION SYSTEMS

Some Central Issues

The increasing global interest in issues related to agricultural sustainability, such as those reflected above, has been discussed in the report of the World Commission on Environment and Development (WCED) and grows out of the concern that many factors are contributing to the degradation of the environment. The intensification of traditional agricultural systems in order to meet growing needs has frequently led to undesirable environmental or ecological consequences, such as soil erosion, salinity, waterlogging and the contamination and over-exploitation of aquifers. It is essential, therefore, that development planners also take into account the related ecological principles when attempting to understand the constraints on increased agricultural productivity.

The technological challenge introduced in the preceding section entails that agricultural production does not merely need to be increased but that its growth needs to be sustained at levels necessary to support a population which is likely to increase well into the 21st Century. Problems of sustainability of agricultural production must therefore feature strongly in any new agenda for agricultural research, and the full weight of modern science and technology must be brought to bear on problems of increasing productivity. Viable alternatives to shifting cultivation and land resting have to be found, and new opportunities in biotechnology exploited for the benefit of developing countries.

However, the prospect of rapid technological progess in understanding the requirements for sustainability could be undermined unless other institutional and policy issues are resolved. For instance, poor farmers have a short-term planning horizon. They are increasingly driven by population pressure and poverty to reduce land resting and extract whatever produce they can from the land to which they have access. This results in a state of land resource which, at best, descends to a low-level equilibrium and, at worst, suffers continuous degradation to the point that it has eventually to be abandoned. Only when incomes rise can short-term expediency give way to long-term planning, and sustainability be taken into account in farmers' decisions. Even so, numerous sociological considerations impinge on the farmers' production function, such as tenurial rights over land and trees. These issues make it difficult for poor farmers to develop adequate considerations for sustainability.

Likewise, governments in developing countries are faced with similar hard choices. Their priorities are often dominated by short-term constraints relating to their current budgets or balance-of-payment situations. Therefore, adequate provision for future needs may be precluded by the urgency of short-term needs, often leaving poorer farming systems to fend for themselves in trying to eke out some form of basic subsistence, until greater national prosperity is achieved (if at all).

There are a number of policy and institutional issues which need to be resolved in this context. A major issue facing project designers and development planners is that of ascertaining who should pay for the cost of the components designed to mitigate problems affecting sustainable agricultural production. For example, the returns from various conservation-based practices are not readily perceived. Even if governments were to shoulder the costs of conservation measures, farmers' involvement from the initial stages of the project would be an essential factor in its implementation and maintenance.

Moreover, conservation-based investments have typically long gestation periods. For instance, the practice of agroforestry could be an essential factor in sustainability, but it requires substantial investment and several years to 'bear fruit' and produce positive net returns on investments. Furthermore, farmers who are constrained by competing demands on their limited resources and labour have to make decisions which may compromise both their current needs and future demands. The task faced by development planners in this regard is daunting. They would need to design an adequate incentive framework for farmers and governments alike to resolve these crucial sustainability issues. Moreover, for each measure proposed, they would have to evaluate who gains, who loses, and who pays for the cost of the measures adopted.

Solutions to problems of sustainability of agricultural production, therefore, do not rest with science and technology alone. They also require vigorous socio-economic research in order to understand the constraints under which the poor farmer operates. Such research would also be required to guide national policies and assist in the formulation of strategies which have a favourable impact on agricultural development. It would have to provide models in which farmers can afford to pursue sustainable agricultural production without having to resort to day-to-day survival strategies which cannot be sustained in the long run. The pursuit of viable technology systems would thus have to reflect concerns about the complex interaction of the myriad social, ecological and economic aspects which are so critical for sustainable agricultural development. The search for such options would entail a very demanding process requiring talent and substantial investment, both in terms of financial resources and time.

While sustainability is a global problem, the environmental degradation affecting sustainability often results from action taken locally. An individual farmer may follow practices which are beneficial to himself but harmful to others. Moreover, disregard for conservation of the environment in one locality, through deforestation or overgrazing, for example, may contribute to problems elsewhere, such as siltation of reservoirs and downstream flooding. Environmental degradation, therefore, cuts across boundaries and social classes so that action taken in one country may adversely affect the natural resources of another.

Shortage of fuel wood is an important sustainability issue affecting both rural and urban populations. FAO has reported that an estimated 250 million people in developing countries live in areas of fuelwood shortage. Because of the continuous depletion of fuelwood resources, a shortfall of about one billion cubic meters has been projected for the year 2000. This shortfall will mainly affect poor rural populations and will increase the demand for alternative sources of fuel. While the shortage of fuelwood is, in itself, an extremely serious problem, its effect on the environment, through the loss of trees and shrubs, is likely to prove devastating unless remedial measures can be instituted. The role of forest cover is especially important to sustainability in protecting the soil and limiting runoff and erosion, while conserving precious water for recharging aquifers and rivers.

Removal of ground cover by overgrazing has some of the same effects as deforestation, leaving land vulnerable to erosion and reducing soil fertility. Widespread deterioration of grasslands is particularly evident in Africa, where livestock numbers are increasing nearly as rapidly as the human population. Overgrazing has led to deterioration of the soil which, in turn, lowers its carrying capacity and leads to an accelerated cycle of degradation, causing further poverty and malnutrition among the rural poor.

The consequences of the various soil degradation processes are the dimunition and destruction of the land's biological production, leading to more serious desertification. Such a process has contributed to the widespread deterioration of ecosystems and has destroyed the biological potential for plant and animal production, at a time when increasing productivity is so necessary. The WCED has indicated that some 20% of the earth's land area suffers slight, moderate, or severe desertification. An additional 6% was classified as severely desertified. Most of these areas are inhabited by poor rural populations, representing IFAD's target groups.

IFAD is conscious of the fact that until solutions are found and implemented, the areas subject to desertification, which are growing rapidly throughout the developing world, will pose serious problems and will eventually threaten the achievement of the Fund's goals in alleviating rural poverty, increasing food production and improving the nutrition of its target groups. This could lead to further deterioration in the sustainability of agricultural production systems in the years ahead. There is an urgent need for IFAD, therefore, to invest as much as possible in the rehabilitation of areas subject to desertification and thus arrest this process before the areas can no longer be rehabilitated. Such investment today would cost much less than if these areas were to be rehabilitated in the years ahead.

The phenomenon of desertification discussed above, affects both arid and humid regions in the world and is born of processes which are both complex and dynamic. Attempts to control desertification usually fail unless this complexity is recognized and both the social causes and the physical symptoms are taken into account. A key factor in ensuring a successful attack on these detrimental processes, however, lies in effective beneficiary involvement at every step. This would strengthen technology by incorporating their indigenous, intimate knowledge of the enviroment they inhabit, while at the same time, ensuring its sustained voluntary maintenance. IFAD's evolving participatory approach to rural development would pave the way toward this goal. Such an approach should be earnestly pursued and reflected in any future research agenda for proposed IFAD support.

IFAD Target Groups - Victims of Unsustainable Development

IFAD recognizes that long-term agricultural sustainability of production cannot be achieved under the projects it supports unless the production system not only avoids further degradation of the environment but also contributes to the restoration and enhancement of its productive capacity. The issues and concepts associated with sustainability bring to the foreground some major issues which have had a cummulatively detrimental impact on the Fund's target groups. The hard choice between preserving a scarce natural resource for future use or exploiting it to meet current survival needs is but one of the myriad dilemmas to be resolved by the resource-poor farmer.

Target groups under IFAD-financed projects respond to this dilemma and the need for intergenerational equity, according to their perceived needs and socio-economic circumstances. However, the problems are extremely complex and the attitudes and considerations of farmers are often at variance with governments, though both opinions are specifically crucial and a trade-off is not always possible. Subsistence farmers with very marginal resources, who represent the majority of

IFAD target groups, usually have a short-term planning horizon and tend to adopt only risk-averse production strategies. They are increasingly driven to exhaust the soil as a result of population pressure, leading to deteriorating productivity, poorer nutrition and a potentially shorter life-span.

Research to promote appropriate technological and socio-economic solutions is lacking, thus limiting the options for investment in many regions of the world, where a substantial population representing IFAD's target group are waiting for solutions. The focus of a suitable research agenda would need to go beyond mere technological considerations, and solutions would have to be shaped both by the potential beneficiaries' perceptions and capacities to absorb new technology, and by the existing, available local resources.

IFAD'S SUPPORT FOR AGRICULTURAL RESEARCH

IFAD's mandate and its "Lending Policies and Criteria" discussed earlier, have provided the guidelines for financing agricultural and rural development projects in developing countries. They have sought to address the central issues which have an impact on the sustainability of agricultural production systems, especially that of vulnerable smallholders. However, IFAD's focus on these issues has not merely been served by its loan portfolio, which frequently includes a concern for strengthening technology transfer systems in order to address smallholder needs, but is also reflected in its increasing support for the generation of appropriate technology.

IFAD's role in financing agricultural research is manifested by a number of grants it has awarded to International Agricultural Research Centres (both CGIAR and non-CGIAR) and regional agricultural research institutes, for conducting research and training in specific areas of importance to smallholder production systems.

These grants have frequently aimed at strengthening national research systems through research networks, as well as reorienting their research agenda to focus on the needs of smallholders. Appropriate technology so generated, has often flowed directly to smallholder beneficiaries through the medium of IFAD's investment projects.

The increasing number of research programmes and investment projects which IFAD has supported, in the domain of conservation-based interventions, has clearly demonstrated the Fund's concern for sustainability and environmental issues. It has sought to support research efforts which have led to the identification of more efficient, environmentally sound production systems within the context of resource poor farmers, and has drawn several conclusions in the light of this experience. For instance, wherever appropriate, IFAD is supporting research on agroforestry: the introduction of multi-purpose trees and shrubs into production systems. While recognizing the importance of this in promoting biological sustainability of the resource base, the Fund has become aware of the need for long-term research in this field. It will seek to continue with its support for agroforestry research, for instance, through the SAFGRAD research programme, and by strengthening the institutional set-up. Other research grants which exemplify this are the ILCA programme on small ruminants in the humid and sub-humid zones which incorporates alley farming, the CATIE farming systems research programme and the ICRAF agroforestry research programme. This latter aims to establish a research network for the semi-arid tropics, and represents a shift in emphasis from the humid and semi-humid ecologies where significant research progress has already been achieved.

Furthermore, the Fund has supported research at IITA, on biological control of cassava pests and at ICIPE to find environmentally safe methods of biologically controlling crop borers. It has supported IFDC in investigating ways and means of improving the efficiency of integrated nitrogen and phosphorous fertilizer use and in identifying cheaper local resources for their production. It has supported IRRI research in the biological fixation of nitrogen through Azolla production and its utilisation in a variety of ecologies by rice-growing small farmers. Water management of irrigation schemes is among other areas related to the rehabilitation, maintenance and enhancement of natural resources, which the Fund has supported through research programmes undertaken by IIMI.

The purpose of this section is to discuss some of the areas of agricultural research concerning smallholders and its organisation, which IFAD has supported in the past. It outlines the aims, impacts and lessons learned through its experience in financing such research. It is also intended here, to highlight the emerging priorities which need to be addressed in a future agenda for research on small farming systems. Moreover, the section implicitly seeks to draw attention to the fact that the technological challenge in this context is formidable and extends beyond the Fund's limited resources. There is a clear need, therefore, to mobilise donor resources in support of the emerging research agenda.

Research on Traditional Food Crops

IFAD's focus on poor small farmers entails the need to focus on traditional food crops. Much of it illustrates the Fund's support for research directed towards the improvement of traditional food-crops and the development of technological innovations which can be adopted by the small farmer with minimum cost. For example, IFAD-supported research conducted by IITA has developed new clones of cassava which show a high resistance to major diseases and have been widely adopted by small farmers in western and central Africa. ICRISAT has developed cold-tolerant, disease-resistant food-sorghums for the high-elevation areas of Latin and Central America and studied their incorporation into local farming systems. CIAT has developed a range of new varieties of phaseolus beans which, while conforming to local preferences for size and colour, also show increased resistance to pests and diseases, are tolerant of water-stress, and result in worthwhile yields with minimum inputs. This programme has also developed improved beans with a climbing habit and appropriate colour to suit the consumers' taste. The beans so developed are, therefore, especially adapted to the needs of small farmers in Latin America.

All of these projects have formed part of the wider research programmes in which the individual centres are engaged. The projects have, therefore, been able to draw extensively on multi-disciplinary support in strategic research and on additional scientific expertise provided from the Centres' core activities. Undoubtedly, without such institutional support, such projects would have been less successful. The improved varieties developed under these research programmes were incorporated in various IFAD-financed development projects for wider diffusion to the majority of IFAD's target groups.

Farming Systems Research

The more progressive agricultural research workers have traditionally extended their experiments to farmers' fields. The wider recognition of the importance of this practice and of the need to take into account the socio- economic circumstances of the small-farm household, has led to the rationalization of this approach under the title of "Farming Systems Research" (FSR) or "research with a farming system perspective". The farming systems approach suggests that in order to give adequate weight to farmers' goals and management strategies, adaptive research is best undertaken through on-farm, farmer- managed experiments. Both the technical and socio- economic factors can then be taken into account and the experiments serve to bridge the gap between research and extension. In the farming systems approach, on-farm experiments are used not only to evaluate innovations, but also to provide feedback to guide research workers at all levels in planning their future work. It therefore adds a further iterative element to the research process by incorporating farmer responses to the strengths and weaknesses of the innovations under test.

IFAD has attached considerable importance to the support of farming systems research, although not all of the research funded under the umbrella of FSR has conformed precisely to the principles enunciated above. Some of it has been mainly concerned with applied research on the components of cropping systems. IFAD's support for research on rice-based cropping systems by IRRI, for example, has been primarily concerned with the integration of improved rice varieties into a multiplicity of cropping systems designed to improve the productivity and stability of food supply under different socio-economic and agro-ecological conditions. While noting significant advances made in this work, the report of the mid-term review also made some important criticism related mainly to the difficulties of centralizing research of this type, especially when it is primarily concerned with farming systems which are inherently location-specific.

In contrast, the decentralized approach to on-farm research on faba beans in the Nile Valley project has met with universal praise. This work has been carried out entirely by national scientists under a networking arrangement with backstopping from ICARDA. It illustrates how, given the local resources and central backstopping, national scientists can respond quickly and effectively to new challenges, while gaining confidence, morale and scientific stature in the process.

To a considerable extent, similar inferences may be drawn from the IFAD-supported SAFGRAD/OAU/STRC project in which African scientists have been introduced to the concepts of farming systems research. Owing to the weakness of the national research systems in many of the countries in the region, this work was not undertaken by national scientists in their own countries, but scientists were recruited from the region to work as part of teams in three countries which were chosen for case studies: Burkina Faso, Benin and Cameroon. Backstopping was provided by IITA and ICRISAT. It was hoped that the experience gained in these countries would have spill-over effects for the other 23 countries in the region.

The project exemplifies several important features. It explicitly includes sustainability as one of the major aims of the improved farming systems and illustrates the principle that ultimate success in research involves both a sustainability and a farming systems perspective. Success is likely to be achieved only if there is a long-term continuity of research effort.

An assessment of the project notes that pioneering projects of this type can and should evolve in such a way as to provide sound technical underpinning for integrated projects in rural development, which can further enhance research and ensure project sustainability. However, returns from investment in research are better safeguarded when this evolutionary process is not left to chance but forms an integral part of the planning from the outset.

The obvious conclusion is that support for adaptive research should always be linked to the realistic prospect of its being progressively integrated into the wider development process. This is best achieved by agreement with the participating countries and, wherever necessary, with assistance from development agencies. Where this principle is not followed, the successful outcome of the research is inevitably undermined. In Sudan, for example, although the initial stages of the faba bean project referred to above have been extremely successful, further progress in expanding faba bean production is now constrained by the lack of availability of the required inputs, such as agricultural chemicals and fuel for irrigation pumps.

In many of the projects it has supported, IFAD has already adopted the principle of closely linking adaptive research to a corresponding development project. The CIP project on potatoes, for example, has been closely linked to development projects funded by IFAD in North and West Africa, while in the Nile Valley project, close linkages between the research and extension services were established from the outset. The Fund might wish to give even greater emphasis to these principles in the future when revising its criteria for the support of adaptive research. Put another way, IFAD should be prepared to continue its support for a given project in adaptive research until it has been successfully incorporated into a wider development project, thus ensuring a critical continuity of effort.

Biological Control

In supporting research on the biological control of cassava pests, IFAD has participated in one of the most outstanding successes in African agriculture in recent years. In large areas of Africa, devastation of cassava caused by the mealy bug has been greatly ameliorated through the discovery, introduction, culturing and release of the parasitoid, Epidinocarsis Lopezi. This success has been achieved through a major collaborative project involving a group of donors "The Sponsoring Group", several international research institutions and a large number of national agricultural research systems in the developing countries, in a concerted effort which became known as "The Africa-wide Biological Control Programme."

IFAD provided the Secretariat for the Sponsoring Group and was the leader in co-ordinating the overall effort. IITA played a leading role among the institutions, which included CIAT, CIBC and ICIPE. National agricultural research systems were actively involved through a networking approach (described below) and, in some instances, their work was incorporated into IFAD development projects. The mid-term review of this project conservatively estimated the benefit-cost ratio as probably being in the order of 20:1, although precise estimation is difficult because of the difficulty in estimating figures for prices and production.

The establishment of E. Lopezi for the control of the cassava mealy bug is an example of the successful implementation of classical biological control in which an accidentally introduced pest is controlled by deliberately introducing predators and parasites from the same region of origin. The aim is to establish a new equilibrium in which the pest and the controlling organisms survive at low population densities. This method is not always successful, however, and attempts to control the cassava green mite using similar methods have not yet met with comparable success. A combination of different approaches is being attempted. This includes the selection of resistent material under different environmental conditions, using a network of national programmes in Eastern and Southern Africa. While it is clear from this and other examples that classical biological control is not always easy to achieve, the biological control of indigenous pests is even less so. Patience and perseverance are required if success is to be achieved.

IFAD's support to ICIPE for research on the biological control of crop borers in sorghum, maize and cowpeas, must be seen in this light. A review of the project stresses the dangers of trying to tackle too much at one time. Since then, the work has concentrated on two species of borer, Chilo Partellus which attacks sorghum and maize, and Maruca Testulalis, which attacks cowpeas. Micro-organisms and insect parasitoids which attack these borers are being studied intensively, methods of mass production developed and the effects of releasing the organisms investigated.

Through a recent technical assistance grant to IITA, IFAD intends to support further research in the intensification of the biological control of pests. This grant will support a major exploration to discover natural enemies of the larger grain borer, Prostephanus Truncatos, which was accidentally introduced to East and Central Africa from Central America, and where it is now threatening stored maize.

Biological control is seen as only one aspect of integrated pest management, and the research is inevitably long-term in nature. Host-plant resistance, which is an important element in the integrated control of these pests, is being developed by other institutes. In this connection there is scope for promoting closer collaboration among all these research institutions working on the pests attacking these crops, in order to exploit the advantages of the interchange of information and genetic material fostered by such an effort.

Livestock

IFAD has supported research on livestock mainly through work by ILCA, but also to some extent through one of the ACSAD projects to which reference has already been made. In relation to IFAD's mandate, livestock is of particular importance in the drier regions where the small farmer may well be largely dependent on ruminants for survival, especially during prolonged periods of drought.

The ILCA project has been concerned with the study of livestock systems in the arid and semi-arid zones of West Africa and with the development of improvements through component research. Extensive data have been accumulated from surveys of livestock systems involving cattle, sheep and goats, from which major constraints to productivity have been identified. Among these, pre-weaning mortality of offspring has been identified as of particular importance and has led to research on nutrition and the availability of fodder, especially towards the end of the dry season. Nutrition is also being studied in relation to the performance of traction animals. Large numbers of leguminous forages and food crops have been screened for possible inclusion in the agro-pastoral sub-system based on millet. Phosphatic fertilizers have also been confirmed as being important. The study of the sub-system is being extended to include leguminous trees and shrubs, especially Acacia Albida as the main tree species. Studies of rangeland vegetation have led to the development of a method for estimating potential range-productivity in the Sahel by using rainfall probability data. The estimates of grazing capacity through satellite imagery is being validated from observations on the ground.

The mid-term review of this project stresses the need for closer collaboration between ILCA and other institutions working on related problems, such as ICRISAT, IITA and ICRAF, particularly in relation to the sharing of knowledge on suitable plants to be included in these complex agro-silvo-pastoral systems. The report also recommends closer collaboration between the staff involved in this project and those involved in the IFAD-supported SAFGRAD project, to which reference has already been made.

As with several of the other projects funded by IFAD, the ILCA project also illustrates the importance of linkages with national development programmes for the achievement of greater impact. Improvement of livestock productivity is closely related to the availability of veterinary services and access to markets for livestock products. These and other aspects of productivity are themselves related to government policies and development strategies. The mid-term review identified several areas where investment in development projects would help in this respect. It is unlikely that ILCA's work in the Sahel will achieve significant impact until these are in place.

Networking

The principle of networking has been applied in many different circumstances. Current thinking has evolved largely from the work of the CGIAR Centres in the establishment of international testing networks for germplasm, the first of which was established by IRRI in 1963. Networks are now widely used for experimental work extending far beyond the germplasm evaluation. With commitment from the individual participants networks can be of great assistance, both in the wider adaptation of technological innovations and in the lateral transfer of knowledge. The establishment of a network does not guarantee success, however, either in the more rapid realization of research goals or in the successful transfer of innovations. In actively promoting this trend towards collaboration and partnership in research through networking arrangements, IFAD has given support to both regional institutions and national systems. In the initial ACSAD project, for example, four countries in West Asia and North Africa were encouraged to collaborate in the breeding and large-scale production of drought- tolerant varieties of wheat and barley. Fertilizer responses were also investigated and seed distribution schemes successfully introduced. Backstopping was provided by ACSAD.

In a more recent phase of this research partnership, six countries are collaborating in a unified and systematic approach to provide greater understanding of the agronomic principles which determine productivity in these drought-prone Mediterranean environments. Various methods of tillage, different combinations of rotations using leguminous crops, the use of fertilizers, and the incorporation of livestock, are all being investigated. Under the leadership of ACSAD scientists, the participating countries have agreed on uniform designs for the experiments, enabling serial analyses of the results to be conducted and wider inferences to be drawn. This collaboration is not only providing many more valuable results than each country could possibly provide individually, but it has also given a new sense of purpose and enthusiasm to the participating staff.

This important principle of collaboration in applied research is further illustrated by IFAD's support for work being carried out on plantains by IITA. In this task a multidisciplinary team of research workers has been built up by drawing on scientists from a range of different institutions and national programmes in Western Africa. The largely informal structure under which this group of scientists works is known as the West Africa Regional Cooperative for Research on Plantains (WARCORP). It is developing improved agronomic practices for plantains, investigating problems of disease control and exploring improved methods of storage and processing.

Socio-economic Research

IFAD has recognized the need to support agronomic and socio-economic research so as to provide new technological bases for increasing the productivity of agricultural resources. In its concern for the welfare of the poorest farmers, IFAD has supported research by IFPRI, designed to analyse the effects of agricultural change on income and nutrition. Among the studies completed so far, one has investigated ways in which the promotion of technological change in methods of rice production in West Africa could assist in alleviating poverty, increasing food consumption, and improving nutrition. The study concluded that many problems of malnutrition, especially those caused by seasonal variations in the availability of food, could be dealt with by improving the incomes of the poorest people and, consequently, their access to food. The study therefore called for policies and development programmes which are broadly based and extend well beyond the promotion of improved farming practices for a single crop. It also pointed out that attention must be given to improving rural infrastructure so as to ensure the adequacy of delivery systems for agricultural inputs and the availability of credit, especially to the poor.

The study also highlighted the important role played by women, not only in relation to the health and nutrition of children, but also in relation to their importance in the production of cash crops. In this connection, the report emphasizes the difficulties faced by women in attempting to improve their productivity without access to credit for the purchase of the necessary inputs such as tools, fertilizers, etc.

The study in Kenya examined the effects of shifting production from maize to sugarcane and also the effects of labour productivity on the incomes and nutrition of traditional farming families in South Western Kenya. It established that while maize usually gave a higher return from a given area of land than sugarcane, the returns to labour were significantly higher in sugarcane production than in maize. The study concluded, however, that although increased income resulting from higher labour productivity in sugar cane was important in contributing to the family's increased food consumption, the effect was somewhat diminished by the damage to health.

Two major conclusions may be drawn from these two studies: the need for the poor, especially women, to have access to credit and the importance of good health. These studies have provided new insights into the linkages between production, income, consumption and nutrition and have reaffirmed the well-known association between health and nutrition. Both require further study.

Working with Research Centres: Some Institutional Issues

The experience gained by IFAD in funding research is also useful in pinpointing some of the advantages and disadvantages of working with different types of research institutions. In general, grants awarded to the International Agricultural Research Centres (IARCs) of the CGIAR have resulted in satisfactory delivery of the required results. This is not to say that reviews of work undertaken by the CGIAR Centres have been free of criticism. On the contrary, the work of the Centres is continually subjected to thorough review, not only by donors such as IFAD, but also by rigorous processes of internal and external review mounted by the Boards of Trustees and by TAC.

Much of the strength of the CGIAR system, however, derives from its willingness to be subjected to external review and by the constructive way in which centre managements respond to criticism. The system must maintain the highest standards of administration and financial management. Especially vital for the future confidence of donors in the System as a whole is the elimination of waste or extravagance.

IFAD's support for regional organizations in general should be seen in part as a reflection of its institution- building policy. Various reviews and evaluations of these institutions point to the fact that they are more likely to experience difficulties in continuity than the international centres. This derives from their vulnerability which is caused by the structural disadvantages inherent in regional research centres and is something that cannot easily be overcome. The member countries are bound to have divided loyalties in which the need for resources in their own countries may well take precedence over their willingness to supply resources to a regional centre. Regional organizations are more likely to succeed if they continue their focus on specific regional issues and problems and thus become centres of scientific excellence in the generation of specific technology. They should also become efficient facilitators of collaborative regional programmes.

IFAD's Research Support : A Sound Investment?

From the examples mentioned in the foregoing sub-sections, it is clear that the scope of the research projects and their content have been consistent with the guidelines and criteria adopted by IFAD for the selection of projects to support. The selection of commodities and topics is closely related to the needs of small farmers in the less-endowed regions. Many of the projects illustrate IFAD's commitment to institution-building and the strengthening of national systems, while simultaneously funding the adaptive research necessary to support its operations. They also illustrate IFAD's concern for sustainability of production and environmental issues.

Nonetheless, looking to the future, the task of improving the well-being of poor small farmers and arresting the environmental degradation presents enormous challenges for science and technology, as well as for the development of the human resources essential for success. IFAD has an important role to play in responding to all of these challenges.

EMERGING OPTIONS

Through its interventions and the various research programmes it has supported in the past, IFAD has acquired a growing volume of evidence which draws attention to the crucial issue of the coming decades - sustainable agricultural development. A major goal for the Fund in the future, therefore, would be to support research which would focus on issues related to sustainability of agricultural production systems. The results of such research could contribute to the impact of its future interventions in terms of better addressing both its target groups and the needs of future generations. An explicit effort to include sustainable agricultural systems as one of the main goals for IFAD's future development strategy is, therefore, a necessary dimension of what is intended to be an efficient attack on rural poverty.

IFAD must, therefore, play a leading role in promoting research for the development of new sustainable technologies. Greater research is required to develop more comprehensive approaches to economic analysis: moreover, methodologies must be found for an evaluation which can capture the dynamics of environmental and socio- economic changes which lie at the heart of understanding agricultural sustainability.

Consequently, IFAD needs to continue its support for research efforts leading to the identification of more efficient production systems which are within the means of resource-poor farmers and which protect their environment from degradation. Efforts initiated by the Fund to explore the potential of complex systems of production, the dynamics of the temporal and spatial relationships between the different species comprising them and the opportunities they present for the control of pests and diseases, are steps in the right direction. These efforts, however, need to be sustained and intensified in order to identify those solutions which can be promoted by the investment projects financed by IFAD during the coming decade.

The next section is intended to provide some directions for a future research agenda. It is by no means an exhaustive list of all issues and areas requiring research attention, but it represents some of the concerns which can be addressed by IFAD directly or through other institutions. The areas of research that require attention are not confined to scientific research alone. For instance, it is recognized that in the drier regions of the world, where rainfall is erratic and water resources limited, science and technology alone will never completely solve problems such as variability of food production. Therefore, the rural poor in these regions must aim at equipping themselves to provide for years of plenty as well as years of scarcity. Socio-economic research must complement technological research in providing solutions which can maximise the benefits of the good years and minimise the suffering in the bad years. The options discussed below should provide some vital directions for future research and development projects which could eventually lead to the required solutions.

Sustainable Agricultural Development: Revisited

As discussed earlier, the goal of sustainable agriculture should be that of meeting the increasing needs and aspirations of an expanding world population without degrading the environment. This goal presents two distinct, but intimately related, challenges: to increase the productivity of land already in use and to preserve natural ecosystems for the benefit of future generations. Both are difficult for poor farmers to achieve.

Traditional systems of shifting cultivation and land resting ultimately mine the soil of essential plant nutrients, unless they are replenished by recycling organic matter or by applying inorganic fertilizers. Conditions in which total recycling of nutrients is possible, seldom exist, if for no other reason than the fact that the generation of income by farmers involves removal from the land of a marketable surplus of produce, taking with it plant nutrients derived from the soil. Moreover, increases in productivity achieved through the use of improved varieties simply aggravate the problem of soil mining, unless accompanied by practices which restore to the soil the nutrients removed at harvest. Even the application of animal manure may deplete nutrients from one area of land to sustain production on another.

In this context, much has been written about the need for low-input production systems, by which is implied the need to develop crop varieties and farming practices which will increase productivity while minimizing the need for external inputs. An aspect requiring further research, however, is how these needs can be easily reconciled with the underlying requirements of sustainability. The current focus of research, as in the past, has been on high potential areas where the main issues have to do mainly with the adaptation of available technology. However, the real challenge for research lies in the marginal areas for which there is a need for the generation of new and viable technologies. In resource-poor areas, research is required to optimize returns from the minimum use of external inputs and to quantify the level of inputs required to ensure that the system is truly sustainable.

In this respect, there are big differences among types of production systems. Well-managed systems of cassava production, for example, can produce large amounts of carbohydrates with relatively little external inputs. In contrast, irrigated systems of multiple cropping with rice or wheat require large amounts of external inputs in order to be sustainable. In the context of the poor farmer, the aim of research and development should be to promote the adoption of balanced production systems in which the requirement for sustainability is fully met. In the longer term, this requirement might well imply an evolution towards the use of progressively increasing external inputs. The stark alternative is to accept further soil degradation and unabated destruction of natural vegetation.

Research must continue to explore ways of improving the efficiency of uptake of applied nutrients, thus reducing loss of nitrogen through leaching or phosphate through fixation. Possible approaches include genetic improvements to the crop plant, encouraging mycorrhizal associations and modifying the formulation of fertilizers.

Continued mining of nutrients from the soil is not the only factor which threatens sustainability. The increasing and inexorable demand for firewood for cooking now poses a threat of equal importance, aggravating pressure on the environment caused by the need for new land. As a result, forests, savannahs and rangelands are being denuded of their natural vegetation at an alarming rate, leading to degradation, erosion, and even desertification in some areas. The incorporation of multipurpose trees and shrubs into production systems has therefore become a major challenge. Research in this area, usually referred to as agroforestry, is relatively new. There is an urgent need to develop methodologies for studying the productive potential of these complex systems in quantitative terms. The dynamics of the temporal and spatial relationships of the species that can be incorporated into them also need to be studied. Although the development of agroforestry systems which can be adopted by small farmers presents a major challenge for research during the coming decade, the prospects for success are encouraging.

Alternatives to Shifting Cultivation

During the past thirty years, a great deal has been learned about the soils of tropical and sub-tropical regions, with respect to maintaining their productivity under arable cropping and preventing erosion. Even in the higher rainfall areas where the natural vegetation is forest, technological solutions permitting the sustainable production of arable crops and, in many areas, livestock, are now available. The problem is that many of these technological solutions are not economically viable under prevailing circumstances. They are therefore beyond the reach of the poor farmer and even if the economic problems were solved, the high level of management skills they require would make their adoption difficult.

Future research must therefore continue to develop new farming practices which are less demanding on the farmer's resources, but meet the requirement for sustainability as well as the needs of the traditional farming household. Various systems of alley-farming developed at IITA have come a long way towards meeting these requirements. The leguminous shrubs in the system fix atmospheric nitrogen, thus reducing the need for applied nitrogenous fertilizer. They provide browse for livestock and wood for fuel and poles. Planted on the contour, they reduce run-off, thereby increasing infiltration and preventing erosion.

More research is needed, however, to quantify the potential output of these systems and their requirements for sustainability over a range of different soil types and rainfall regimes. The range of possible multi-purpose shrubs needs further evaluation, particularly in relation to their acceptability to farmers and problems of pest control.

With possibilities for eventual reduction in the requirement for land preparation and weed control, this system of alley-farming promises to provide a major breakthrough in tropical agriculture in the coming decade. Its widespread adoption by small farmers, however, will depend on further technological refinements such as those outlined above, as well as on a major effort in adaptive, on-farm research; with particular attention to the socio-economic constraints to its adoption by small farmers.

Increasing Productivity in the Drier Regions

The key to increasing productivity in the drier regions lies both in the more efficient use of rainfall and in the better integration of crops and livestock. There is scope for a great deal of further research on both. Research has improved the understanding of the rainfall distribution in the drier areas and has enabled crops, cropping sequences and grazing regimes to be tailored more precisely to the expectations of available soil moisture. For example, early-maturing maize varieties have been bred successfully in areas of unreliable rainfall, where there are short, but relatively reliable, mid-season periods of rain. Continued research in agrometeorology may eventually lead to greater predictability of seasonal weather, making it possible to advise farmers to adjust their sowing dates or to alert entire communities to the possibility of drought. This is a difficult area of research, however, where progress is unlikely to be fast or spectacular, but the dividends for success may prove critical for rainfed areas.

As the scarcity of arable land increases, irrigation enables farmers to increase production on existing land. Some of this expansion in irrigated areas has been achieved at the expense of using fossil water or of the overdrafting of rechargeable acquifers. Both practices are unsustainable. In addition, chemical and biological pollution are making water unsuitable for irrigation in some localities. These problems are aggravated by the fact that a great deal of irrigation water is used inefficiently: much more water than necessary is frequently transported and supplied to crops. Poor irrigation practices result in severe problems of land degradation through water logging and salination and are estimated to render unproductive some 1-1.5 million hectares of good cropland annually. Indeed, in many areas of the world, waterclogging and salinization threaten to diminish the very gains that expensive new irrigation projects are intended to provide. IFAD is already supporting several research activities to address these issues.

Research must also continue on methods of water-harvesting and on the management of arable land to conserve moisture, prevent erosion and improve productivity. Breeding plants for drought tolerance, increased water-use, and stability of yield under these conditions is important. They could never, on their own, provide the reliability of yields achievable under irrigation, particularly where improved management of irrigation systems is practised or in regions with more favourable rainfall and distribution. Consequently, research for the drier areas must explore every possible means of providing supplementary water during periods of mid-season drought.

IFAD recognizes that research for the development of effective methods of weed control could probably provide a greater contribution to crop productivity in developing countries than new varieties of crops. Various parasitic weeds belonging to the genera Striga and Orabanche are also increasing in importance as crops spread into new areas, rotations are shortened and both organic matter and nitrogen in the soil are depleted. Such innovative methods of weed control present new challenges to science, and, if not carefully monitored, they too could pose an additional threat to sustainability.

Livestock also play a very important role as a buffer to the effects of unpredictable fluctuations in the patterns of rainfall for small farmers as well as for pastoralists and agro-pastoralists. Research on crop-livestock interaction, (and on the management of livestock and the forage that sustains them), will continue to be of great importance over a wide range of agro-ecological conditions and socio-economic circumstances, especially in the drier regions.

Biotechnology

Each year, pests destroy enough of man's food and fiber crops to feed and clothe over 20% of the world's population. In order to reduce these losses, researchers have spent decades developing chemical pesticides and breeding resistant crops. Through biotechnology, crops have now been genetically engineered to resist major insect pests by introducing into their genomes a bacterial gene which produces a natural, non-toxic and biodegradable protein insecticide. The extent to which recent advances in biotechnology, particularly those in molecular genetics, will contribute to solving agricultural problems in developing countries during the coming decade, is a topic which gives rise to considerable controversy. Part of the problem lies in the fact that many of those involved in basic research in the biological sciences have not been exposed to the real problems of tropical agriculture in developing countries.

Biotechnology is not a new area of investigation; in fact, during the past thirty years, many new techniques stemming from biotechnology have been incorporated into plant breeding programmes as a matter of routine. These include tissue culture, micro-propagation, embryo-rescue, anther culture, soma-clonal variation and virus indexing, to name but a few. Several of these techniques were supported by IFAD's technical assistance programme. Other techniques emerging from molecular biology have either been adopted by plant breeders already or are in the pipeline and will be widely adopted in the coming years. Recently completed field tests on transgenic tomato plants bearing these new traits demonstrated these high levels of efficacy for each gene. Techniques for transferring genetic resistance to tobacco mosaic virus infection in transgenic plants, as well as in identifying the cellular and molecular mechanisms responsible for the trait, have been developed recently. Many of these techniques have the effect of accelerating existing plant breeding procedures so that the rate of producing new varieties will also accelerate. Other innovations relate to increasing the taxonomic distance over which gene exchange can occur.

Although there is no theoretical reason to prevent genes from any existing organism from being transfered to any other existing organism, we are a long way from doing so. Biotechnology research may make it possible to modify existing genes or even to construct new ones by recombining known sequences of DNA. Even when all these new possibilities are within the grasp of the plant breeder, however, it would be very difficult to predict what the benefits and dangers will be. Effecting improvements to living organisms through genetic manipulation at the molecular level is extremely difficult because of the complexity of the physiological processes determining their performance. In all but a few instances, it is unlikely that the delicate balance of form and function, built up through millions of years of evolution, will be easily improved upon. This is especially true of complex characteristics such as yield and drought tolerance. However, recent success has been reported in transferring salt tolerance to wheat, which allows increased production under conditions of moderate soil salinity or permits the use of mildly saline water for irrigation. Such technology makes possible the development of a whole new range of salt tolerant varieties of wheat.

The coming decade may well see some significant breakthroughs in less complex characters which could be of considerable importance for developing countries and the small farmer. Among the most likely are improvements in weed control through the use of more effective herbicides and the development of novel forms of resistance to pests and diseases. Recent advances in cell biology and molecular genetics offer the prospect of either selecting crop plants that are resistant to universal herbicides or of incorporating resistance genes in them through a process of genetic transformation. Both approaches could result in combining the cultivar and the herbicide into a single, complementary package.

There might well be possibilities in the future, therefore, of producing herbicides that are within the reach of the small farmer, universally effective against all weeds, easy to apply, and environmentally safe in that they could be made to be rapidly biodegradable. Given the time and resources, major advances of this type are feasible and would radically transform agronomic practices in the tropics, where weeds constitute one of the greatest constraints to the productivity of both crops and labour. Moreover, safe and effective herbicides would make zero-tillage a more attractive prospect, largely removing the need for primary cultivation.

Other developments which are within sight but for only a few species so far, are the genetic engineering of novel forms of resistance to virus diseases and the incorporation of genes which will cause plants to produce their own insecticides. For example, the gene coding for an insect toxin produced by the bacterium Bacillus Thuringiensis has already been successfully transferred to the tobacco plant. Experiments have shown that on some transformed plants, tobacco hornworm larvae stop feeding within 18 hours and are killed within three days.

The general application of this type of genetic engineering, however, must await the availability of workable techniques for transformation. These techniques have not yet been developed for the majority of crop plants, in spite of a very considerable research effort directed towards this end. In any case, before genetically transformed foodcrops could be released for general cultivation, an extensive period of evaluation would be necessary to ensure that there were no adverse effects on the consumer or on the environment. In the case of plants engineered to produce insecticides, for example, it would be necessary to investigate the effects on other organisms, such as on beneficial insects. Requirements such as these would increase research cost and lengthen the time required before new varieties could be released.

These are just a few examples of the many new techniques emerging from the biological sciences which have wide application to the solution of problems in crop and animal productivity. They are typical, however, in their clear demonstration that although biotechnology offers some exciting new possibilities, it is unlikely to produce quick or easy solutions to the major problems facing sustainable agriculture in developing countries. Indeed, it has nothing at all to contribute to some of these problems.

On the other hand, driven by the profit motive, private-sector organizations are already in the forefront of this type of research. The initial cost of the products, if successfully developed, might well be beyond the means of the small farmer in developing countries. The forces of commercial competition are such, however, that prices are unlikely to remain indefinitely at levels which are totally beyond the reach of the small farmer. Moreover, possibilities for collaboration between the private sector and international centres in the joint production of packages of this type need not be ruled out. As the linkages keep increasing between the centers and the advanced institutions and the private sector throughout the world, progessively wider scientific contributions will be brought to bear on the problems of crop and livestock productivity in developing countries.

Research on Socio-economic Factors

One of the most serious limitations to sustainable agricultural development relates to the focus of domestic policy in many developing countries. Often both national and local governments accord low priority to the agricultural sector. Budgetary allocations to this sector are frequently eclipsed by those accorded to other sectors, there is a general lack of rural industry and, subsequently, few income-generating options are open to the rural poor. Lack of political support is further aggravated by discontinuity in government administration, often associated with political instability. These deficiencies in financial, political and administrative support present formidable obstacles to agricultural development and the achievement of sustainable agricultural production.

Agricultural prosperity is a prerequisite for rural industry but lack of rural infrastructural investments throttles attempts to achieve this. Furthermore, a weak infrastructure is a major constraint to the delivery of inputs and the "evacuation" of farm products to markets. Technical and Economic Policy Research needs to focus on alleviating these constraints and to assist in providing a suitable direction to development policy.

In addition, research on fiscal policy should provide a viable incentive framework which encourages farmers to pursue production systems that incorporate a concern for the conservation of the natural resource base. It also needs to address the problem of how conservation-based investments of time and labour on the part of the farmer can be made more attractive. At the same time, it should propose a fiscal disincentive framework which would discourage short-sighted production strategies (which promise immediate returns to farmers, but which deplete their resource-base in the long run). Such research can also provide the basis for a strategy to encourage extensive production in favourable areas, thus reducing the pressure for increased production in the more fragile environments.

In many circumstances, the achievement of sustainable production will also require the increased use of purchased inputs such as seeds, fertilizers, pesticides, implements and farm equipment. Research must also provide solutions to eliminate some of the major constraints to their availability, at reasonable prices for resource-poor farmers. A poor distribution system causes supply bottlenecks, and, in the process, resource-poor areas are confronted with higher prices than those prevailing in high potential areas. Consequently, research needs to recognize and respond accordingly to the inadequacies of price policies which favour surplus regions.

IFAD has established that credit can be an important vehicle for the alleviation of poverty in rural areas. The experience gained through the first decade of IFAD's operations points to the need to continue the search for innovative approaches to credit delivery and recovery, to the mobilization of rural savings and to the promotion of credit accessability to special groups such as rural women and tribal populations. Further research must be aimed at identifying appropriate mechanisms and innovative arrangements for the institutional financing of small farmers.

Although considerable progress has been made in some Asian and Latin American countries in the development of efficient national agricultural research and extension systems, its performance is, however, still weak. This was confirmed by the findings of the Seminar organized by IFAD on "The Generation and Transfer of Technology to Poor Small Farmers: Issues and Options", which was held in Korea in June 1988. One of the findings of this conference was that research on the development of improved systems for technology transfer and innovative linkages of their different components requires greater attention. This is essential if poor small farmers are to receive the desired impact of appropriate technology.

Institutional interventions are often impeded by conservative national policies. One example is the lack of initiative to address issues of land tenure. On the other hand, a fragmented agrarian structure has impeded the adoption of conventionally and traditionally sound technological options. For instance, it has led to diseconomies in the use of resources and has restricted the adoption of technology packages focussed on entire watershed communities (in rainfed areas) and command area communities (in irrigated areas). Research will have to resolve the issues of how these externalities can be internalised in the context of surmounting the agrarian structural barriers. Furthermore, socio-economic research on issues related to land tenure may contribute significantly to improved sustainability. For example, customary land tenure, usufruct rights and the rationalisation of communal land use to ensure its conservation constitute important research issues for sustainable development. Socio-economic research on such issues will contribute to a better understanding of the institutional mechanisms, policies and approaches required to address development issues in different regions. For instance, it will need to address such issues as conflicting demands for land use by pastoralists and cultivators in the Sahel. The issues to be resolved here are more closely related to the sociological and political context of pastoral development (as well as the economic viability of traditional pastoral husbandry) than to the continual search for technological options for increasing productivity under Sahelian conditions.

The fact that increased agricultural production alone does not necessarily account for household food security and adequate nutrition, underscores the need for further research on this vital subject. As mentioned earlier, IFAD has already supported research (through studies by IFPRI) on the income and nutritional impact of agricultural production. The insights provided by these studies should form the basis for further research to address how the impact of the linkages between various parameters of production, income, consumption and health can be made positive and sustainable for small farmers.

Although health problems *per se* lie outside the immediate field of IFAD's operations, their importance is such that other linkages are clearly difficult to evaluate unless the health and sanitation problem is tackled at the same time. Therefore, that future research should be designed to elucidate the socio-economic effects of agricultural change at the household level and be closely linked to integrated rural development projects in which health problems are also being tackled. Work along these lines might well require collaboration between IFAD and other development agencies more directly concerned with health problems.

Conventionally, research has focussed solely on increasing productivity and cost effectiveness. While these are critical issues, such research has been oblivious to the broader issues of providing gainful employment and absorbing both skilled and unskilled "surplus" labour in rural areas. Research has also often ignored the many advantages of traditional agriculture over commercial agriculture. A more holistic, Technical Systems research approach can provide technologies which reconcile the need for higher productivity with that of maximizing participation, productive employment and the generation of income, without displacing labour. This approach may be the key to alleviating hunger and poverty.

In the light if these facts, a reassessment of the Farming Systems Research (FSR) shows that it has not led to the improvement of the sustainability of major farming systems in any significant way. A modified approach currently emerging is the Rural Systems Research (RSR) programme which is designed to be truly integrative; not only in terms of crop and animal husbandry, but also in the context of strengthening growth linkages among the primary (farms), secondary (Agro-industries and rural industries) and tertiary (services) sectors of the rural economy. The focus of systems research will be broadened, in the process, to include off-farm activities, which are becoming increasingly important in sustainable development. Such an approach would reorient FSR in a way that would lead to a desirable mix of methodologies involving a combination of adaptive and strategic research programmes designed to start with small farmer problems and work towards scientific approaches to solving them.

MEETING THE CHALLENGES

All the foregoing examples of future challenges for research to meet the needs of the poor farmer illustrate the complexity of the problems involved and the importance of harnessing wide support in order to solve them. Research by individuals or even by large teams, however strongly motivated, will not be sufficient if these specialists are forced to work in relative isolation. An exchange of knowledge and ideas must be actively encouraged at all levels. Collaboration among individuals must be reinforced by collaboration among institutions so as to provide the flow of new knowledge, the continuity of effort, and the institutional memory without which the continuum of research cannot be maintained.

All research carries with it an opportunity cost represented by an alternative approach to finding a solution to the same problem. There is no doubt that the opportunities for further advancement through the use of well established methods of agricultural research have by no means been exhausted. Consequently, there is little to be gained from using scarce resources for research which, although intellectually exciting, have a high risk of not providing worth-while pay-offs in terms of practical applications to the problems of the poor farmer. The difficulty sometimes lies in defining "high-risk" or in determining which approach is likely to be the most cost-effective. There are dangers both in exaggerating the likely benefits from frontier technology on the one hand, and of being unreceptive to new ideas (and not exploiting real opportunities) on the other.

Donors and development agencies have heavy responsibilities in these respects. They must recognize that research cannot succeed unless it is adequately funded and unless the flow of new knowledge is continuously ensured throughout all levels of research. Their commitment, therefore, must be long-term and should cover all levels of research, while making provision for collaboration among individuals and institutions.

In tackling the needs of the poorest farmers in the least-advantaged areas, IFAD has set itself some difficult tasks. Increasing the productivity of marginal areas and counteracting environmental degradation requires the urgent support of innovative research. We need a far greater understanding of the physiological limits to the productivity of crops and livestock in these areas; we need to explore the opportunities created by modern science and technology for maximizing returns from scarce resources; and we need more knowledge of the agro-ecological and socio-economic principles which lead to sustainable agricultural production in these fragile environments. Without major breakthroughs in science and technology relating to these difficult issues, IFAD will be unable to succeed in those areas where success is needed most and will ultimately fail to meet the challenges it has set out to achieve.

ABBREVIATIONS AND ACRONYMS

ACSAD	Arab Centre for the Study of Arid Zones and Dry Lands
CATIE	Tropical Agricultural Research and Training Centre
CGIAR	Consultative Group on International Agricultural Research
CIAT	International Centre for Tropical Agriculture
CIBC	Commonwealth Institute of Biological Control
CIP	International Potato Centre
FAO	Food and Agriculture Oganization of the United Nations
FSR	Farming Systems Research
IARC	International Agricultural Research Centre
ICARDA	International Centre for Agricultural Research in Dry Areas
IFDC	International Fertilizer Development Centre
ICIPE	International Centre for Insect Physiology and Ecology
ICLARM	International Centre for Living Aquatic Resources Management
ICRAF	International Council for Research in Agro-Forestry
ICRISAT	International Crops Research Institute for the Semi-Arid Tropics
IFAD	International Fund for Agricultural Development
IFPRI	International Food Policy Research Institute
IIED	International Institute for Environment and Development
IIMI	International Irrigation Management Institute
IITA	International Institute for Tropical Agriculture
ILCA	International Livestock Centre for Africa
IRRI	International Rice Research Institute
OAU/STRC	Organization of African Unity/Scientific, Technical and Research Commission
RSR	Rural Systems Research
SAFGRAD	Semi-Arid Food Grains Research and Development
TAC	Technical Advisory Committee of CGIAR
WARDA	West African Rice Development Association
WARCORP	West Africa Regional Cooperative for Research on Plantains
WCED	World Commission on Environment and Development

BIBLIOGRAPHY

Arnold, M.H., An External Review of IFAD Support for Agricultural Research and Related Training Activities, IFAD, 1988.

Consultative Group for International Agricultural Research (CGIAR), 1968-1987, Annual Report, 1987.

Consultative Group for International Agricultural Research (CGIAR), Impact Study, 1986.

Conway, G.R. and E.B. Barbier, Sustainable Agriculture for Development, Sustainable Agriculture Programme, IIED, London, July 1988. Food and Agriculture Organization of the United Nations (FAO), Dynamics of Rural Poverty, 1986.

Food and Agriculture Organization of the United Nations (FAO), Agriculture: Toward 2000, Rome, 1981.

International Fund for Agricultural Development (IFAD), Annual Report, Rome, 1988.

International Fund for Agricultural Development (IFAD), Comparative Review of Agricultural Research and Extension in IFAD Projects, Draft Report, Rome, February 1989.

International Fund for Agricultural Development (IFAD), Credit Study, Rome, 1988.

International Fund for Agricultural Development (IFAD), Environment, Sustainable Development and the Role of Small Farmers: Issues and Options, presented to the International Consultation on Environment, "Sustainable Development and the Role of Small Farmers", IFAD, Rome, October 1988.

International Fund for Agricultural Development (IFAD), Progress Report on IFAD's Approach to Environment, Sustainable Development and the Role of Small Farmers, presented to the Twelfth IFAD Governing Council, Rome, January 1989.

International Fund for Agricultural Development (IFAD), Generation and Transfer of Technology to Poor Small Farmers: Issues and Options, Rome, 1988 (in print).

International Fund for Agricultural Development (IFAD), Lending Policies and Criteria, Rome, 1988.

International Fund for Agricultural Development (IFAD), Strengthening National Wheat and Rice Research Systems, Rome, 1987.

Kesseba, A.M., J.R. Pitblado and A.P. Uriyo, Trends of Soil Classification in Tanzania: The Experimental Use of the 7th Approximation, The Journal of Soil Science, Vol. 23, No.2, Clarendon Press, Oxford, 1972.

Kesseba, A.M., On the Future of Soil Survey and Land Use Planning for Agricultural Development in Tanzania, In Proceedings of the 1st Conference on Land Use in Tanzania, University Press, 1970.

Postel, S., Conserving Water: The Untapped Alternative, Worldwatch Paper 67, 1985.

Steppler, H.A. and P.K.R. Nair, Agroforestry: A Decade of Development, International Council for Research in Agroforestry (ICRAF), 1987.

Swaminathan, M.A., Sustainable Development Systems for Small Farmers: Issues and Options, presented to the International Consultation on Environment, "Sustainable Development and the Role of Small Farmers", IFAD, Rome October 1988.

Technical Advisory Committee, CGIAR, TAC Review of CGIAR Priorities and Future Strategies, 1985.

CHAPTER TWO

GENERATION AND TRANSFER OF TECHNOLOGY FOR POOR, SMALL FARMERS

Martin E. Pineiro 1/

Small farmers are an important segment of Latin American society. They represent the largest group in the agricultural sector and produce a significant portion of the region's basic foods, as well as a wide range of other commodities.

The small-farm sector is of key importance in the economic crisis confronting Latin American countries. With a significant share of economic resources located in the agricultural sector, reactivating the region's economies essentially means reactivating the agricultural sector. That can be done only if small farmers are fully incorporated into the development effort.

Small farmers should be viewed not only in the traditional perspective of providers of inexpensive food, but also as key players in the effort to diversify agricultural exports. Their sheer numbers illustrate their strategic importance to economic recovery based on the reactivation of agriculture.

Agricultural technology has played, and will continue to play, a substantial role in relation to the welfare of the small-farm sector. However, the issue of generation and transfer of technology has to be

---

1/ Director General, Inter-American Institute for Cooperation on Agriculture (IICA), P.O. Box 55-2200 Coronado, Costa Rica

approached with caution and realism. In most cases, the lack of availability of, or access to, technology is not the central element contributing to the poverty of small farmers. Access to resources and other services is much more likely to be the major contributor. But even if improved technology alone will not solve the problem of rural poverty, it is undeniable that productive land use is critical to "campesino" welfare and appropriate technology is essential for this to improve. Past poor technological performance of the small-farm sector can be blamed in part on a piecemeal approach that did not consider the complexity of the situation. Often technology was, at best, a second or third option for solving problems. In addition, the institutions that generate technology failed to offer the small farmers viable options consistent with their needs and resources.

In this paper, I will address the issues related to technology in the small-farm sector, look at some of the shortcomings of past approaches and offer suggestions for future actions. Furthermore, I will:

- describe the main characteristics, importance and potential of the small-farm sector,
- define the technological problems to be confronted,
- review the principal means of generating and transferring resources to the small farmer, and
- summarize the components of a strategy for research and technology transfer to small farmers.

I do this with great care and humility because of the technical and social complexities of the issues involved. But I also do it with conviction because of the moral and political dimensions of the problem. Better technology should not be considered a deus-ex-machina that will miraculously lift small farmers out of the morass of restraints they currently experience. Yet if a real linkage is made between technology and the population that will use it, there is reason to be optimistic.

Basic considerations of equality of opportunity make the elimination of agricultural poverty a top priority in agricultural development. Most important, a more equitable participation of all sectors in general, and small farmers in particular, in the distribution of the benefits of agricultural growth is necessary for the consolidation of the nascent democracies of Latin America.

The present economic crisis confronting Latin American countries implies public budget reductions that directly impinge on the quantity and quality of services to the small-farm sector. Public research institutions, already suffering from a budget squeeze during the late 1970s and early 1980s, have only sufficient resources to cover salaries. In this situation, a review of technology generation and transfer strategies for the small-farm sector becomes imperative.

THE SMALL-FARM SECTOR

Characteristics

It is a Herculean problem to identify all the characteristics of the small-farm sector and quantify the impact it has on the economies and societies of Latin America. As mentioned in a previous paper (Pineiro et al. 1981), the difficulty lies in the methodological problems of defining and circumscribing a highly complex and diversified subject -- one that has a built-in tendency to be vulnerable to external stimuli that provoke change (Murmis and Cucullu 1980).

Looking at some of the characteristics that are common to the individuals that make up the sector, we (Pineiro et al. 1981) defined small farms as those that have a small hired labour component and rudimentary means of production -- that is, those units in which the process of production is achieved, fundamentally, through a combination of land and family labour.

That simple definition takes into account the main feature common to different small-scale production systems -- a strong interaction between the land and the family unit. In these units, the main resource is the labour provided by the farmer and his family; the hiring of outside labour is relatively infrequent (Table 2.1). When there is a shortage of manpower as a result of increased demand during certain stages in the production cycle (planting during the rainy season, harvesting, handling livestock), it is usually dealt with through a system of sharing and exchanging labour with neighbouring farms. Income from the sale of manpower outside the unit is relatively low. However, when the production system involutes and the unit is subdivided, wage labour outside the farm begins to account for a significant part of the family income (Table 2.2).

In addition, a high percentage of produce is consumed on the farm. This, plus the fact that the family needs some cash income to cover basic needs -- clothing and home construction materials, for example -- results in a relatively diversified, intensive production system where short-cycle crops and small animals are predominant. Thus, the farmer is able to meet his basic needs for food and periodically obtain cash. Such a system, however, is likely to decrease soil fertility and cause erosion.

Small-farmer incomes are low and highly variable because of the effects of external circumstances -- weather and prices, for example --

and savings are also low or absent. Therefore, there is little or no growth. Available data on small production units in Latin America bring to light certain features that should be taken into account when proposing technological change.

In the first place, it is important to understand that the problems of the small-farm sector are complex and that the category "small farmer" actually covers a variety of situations. There are small farmers who own their means of production and therefore have considerable stability as regards their place of work. Other small farmers do not own land and therefore are much less stable. This - to take only one example from a complex situation - means that very different factors are involved in farmer decisions regarding choice of production activities.

Likewise, there are small truck farmers who sell their surpluses on the market and maintain frequent contact with urban centers. This provides increased social exchange and influences the farmers' frame of reference for decision-making. Conversely, there are small farmers who work as seminomadic shepherds in isolated mountain regions. This gives them a different frame of reference - one influenced by natural phenomena.

In the second place, the small-farm sector is a diversified one, contrary to that of larger-scale commercial farmers, whose behaviour is much more uniform because their decison-making is chiefly influenced by the market. In the case of small farmers, they are significantly different among countries, among regions of a country, and even among producers of different crops within the same region. In any event, we are dealing with very different situations within the overall category of small-scale units.

Relative Importance and Trends

Data for the early 1970s indicate that small farms represented a significant share of the total number of production units in several Latin American countries, that they occupied a small percentage of farmed land and that there was no clear trend as regards their average size and relative importance to the region as a whole. Information on nine selected countries shows that, during the period between 1950 and 1970, the market share of small farms in four countries had risen between 10% and 20% (Costa Rica +17.6%, Honduras +10.4%, Brazil +14.4%, Chile +11.5%); in two countries the increase was less than 10% (El Salvador +5.9%, Colombia +4.6%), while in the remaining countries, the share of small farms had decreased by as much as 20% (Mexico -21.2%, Guatemala -1%, Venezuela -19.1%). During the same period, in the same group of countries, the average size of farms included in the analysis was between 1.1 and 2.5 hectares. Farm size had increased between 9% and 20% in three of the countries (El Salvador +9%, Honduras +9%, Chile +21%), had remained constant in another three (Costa Rica, Guatemala and Venezuela) and had decreased between 9% and 19% in the remaining countries (Mexico -19%, Colombia -9%, Brazil -12%).

Recent studies (CEPAL-FAO 1986; FAO 1987) stress that small rural production units are key elements in the Latin American social and economic scene. The studies show the economic importance of those production units and state that, as regards the main crops, half the cultivated area is practically in the hands of small farmers, and that a significant proportion of production is also in their hands. This is true for staples or export commodities (Tables 2.3 and 2.4).

The Latin American rural population is growing in absolute terms and will probably increase from 126 million to 134 million between 1985 and 2000. In Brazil, for example, between 1960 and 1980, the agricultural population working on farms of less than 50 hectares rose by 39.4%, while the population of farms having an area of more than 50 hectares fell by 14%. This indicates that the rural population is increasingly concentrated on the smaller farms.

Although recent data show an overall growth in the number of small farms in general, questions remain regarding trend and magnitude. While in Colombia there has been a decrease in the total number of small farms, and in Brazil they tripled between 940 and 1980, Jamaican figures for the period 1954-1979 show a clear difference between farms of less than 5 hectares, which grew by 8%, and units of between 5 and 25 hectares, which show a considerable decrease. This latter phenomenon would appear to indicate that there is a clear trend towards subdivision and conversion of small farms into minifundia (CEPAL-FAO 1986).

The emergence of new farmers and the movement of others away from farming appear to be taking place simultaneously in the region, in what seems to be an irregular flux. These moves in different directions are probably a result of the expansion of the agricultural frontier, by which new production units are created in some areas while farms that are already small are subdivided into even smaller ones.

SMALL FARM TECHNOLOGY

A System in Precarious Equilibrium

The small rural production unit is a unique and complex system resulting from the interaction of the available factors (basically land and family labour) and of a set of limiting external factors, which condition the level of efficiency in production.

The scarce resources available and, in particular, certain external factors (pricing policies, market, opportunity cost of labour, changes in weather, and others) make the system a fragile and highly unstable one. The process of differentiation, which results from the action of external factors on this precariously balanced system, can lead it to either of the two situations seen in Fig. 2.1. The behaviour of production units based on the use of family labour and rudimentary means of production seems to fall within this pattern.

There is not enough reliable information to adequately identify and quantify the situations of the various types of small farms, and thus it is not yet possible to determine the prevailing trend in Latin America. The studies referred to, however, seem to show not only that small farms in some areas are subject to external factors that create situations that will cause a gradual breakdown, but also that this involutionary process is having a negative impact on more developed production units. This is confirmed by figures showing an increase in the number of minifundia and a decrease in the average area of farms.

A Conditional Solution

The above brings us to the crucial question: Is it possible, through regulation of technological factors, to make a positive impact on the precarious balance of the production system, bringing about an evolutionary differentiation that would gradually and consistently strengthen its productive capacity?

Before this question can be answered, it is necessary to consider some of the features of the small-farm system that will affect any technological approach to the problem of its precarious balance and possible involution.

Rejection of Additional Investment and Increases in Traditional Operating Expenses

Small farms are operated within a context of scarce resources and constant concern with meeting the basic needs of the farmer's family. Since income is barely adequate -- and sometimes inadequate -- to meet basic needs (housing, food, clothing and inputs required periodically to keep the production system going), small farmers are particularly reluctant to take risks. Because they need a constant income to cover these basic needs, they instinctively make decisions that are based more on an assessment of gross profits than on any potential cost-benefit ratio. They are much more likely to try to reduce costs in order to ensure steady income to cover their needs than to try to increase their productive efficiency by using external inputs. A clear example of this is the fact that most small farmers keep part of their grain harvest to be used as seed for the next planting, rather than purchase improved seed from outside their own farms.

Inertia of the System

The production system followed by small farmers is a rational one because it is the best they can manage, considering the resources they have and the constraints within which they work -- the latter including their own limited skills and knowledge. This, when added to their special relationship with the land, their habits, and other factors mentioned above, means that they will have a natural and logical resistance to any proposal involving significant changes in the production system with which they are familiar.

Scarcity of Resources

Basically, small farms have very limited land area. Observation of the main regions where small-farm economies tend to concentrate suggests that they are usually associated with adverse environmental conditions (Renard et al. 1983), and what is available has limited productivity, given a relatively constant amount of family labour and the farmer's particular skill in combining and managing available resources. These elements are the basic components of the production system and are normally the only ones the farmer can use.

Different Agroproduction Situations

The situation with respect to demand for technology differs greatly within the universe of small-scale production systems. This is a result of the differences in environmental conditions and the different ways in which the factors of production are combined, all of which create different technological limitations. Thus, it is important to ensure that the technology offered is consistent with the particular situation of the small farmer concerned.

EXPERIENCES IN TECHNOLOGY GENERATION AND TRANSFER: A MISMATCH BETWEEN GOALS AND RESULTS

Performance of Agricultural Extension Systems

The Latin American countries have long been concerned with the need to generate and transfer technology to improve the level of production, the income and the living standards of small farmers. In the mid 1940s, countries of the region began to create organizations designed to promote the adoption of improved technologies for agricultural production. Those extension organizations were almost all identical in that their conceptual framework and methods were based on the agricultural extension system developed in the United States. The agricultural extension strategy was geared mainly towards the small farmer and was aimed at helping the rural population improve their living standards and at raising the educational and social standards of rural life (Maunder 1973).

The way in which technology was transferred to farmers varied among countries, largely in order to adjust the work of the agricultural extension service to the limited resources available to them. Priority was generally given to technical assistance (either to groups or to individuals) communications, training by demonstration, and supervised credit. The common denominator of all programmes was the assumption that there was a natural and inevitable articulation between the research process and its results, on the one hand, and the process of transferring know-how and its adoption by the farmer, on the other (Trigo et al. 1982).

It is useful to review the reports of the different international conferences held in Latin America between the 1960s and the 1980s on the subject of agricultural extension, in order to get an overall idea of how the countries of the region perceived their own experiences (Indarte 1982). It becomes clear from those reports that, despite the considerable effort put into a large number of experiments in different countries, including some recent ones based on the so-called training-and-visit (T&V) system, the effort to transfer technology to small farmers has not been satisfactory. Each experiment attempted to place the technology generated in the research centres within reach of the farmer. The assumption that there was a natural and inevitable articulation between generation and transfer provides an interesting basis for analyzing some of the causes of the unsatisfactory results.

With few exceptions, research and extension have operated more or less independently. Research and extension activities were compartmentalized, and relations between the two were more circumstantial or personal than institutional and permanent.

This problem was detected at an early stage and several attempts were made to correct it. These ranged from placing the research and extension services in the same unit and under the same chain of command, to - in one extreme case - proposing that an extensionist should head the research service and that a researcher should head the extension service. Obviously, this was not successful, and problems continued to be identified as "research problems" or "extension problems" and to be considered "linked, yet separate" (Ruttan 1987).

The idea that the extensionist is the bridge between research and extension has actually worked contrary to expectations, inasmuch as it has led researchers to generate technological products without taking responsibility for their adoption, their usefulness or their capacity to really help the small farmer. Extensionists, in turn, have received finished products without having had any responsibility for their generation and, consequently, for their adoptability, usefulness and capacity to serve farmers.

Inappropriate Technology

The strategy followed in designing and implementing research and technology transfer projects has been based more on the supply of technological information than on the prior consideration of limitations at the farm level. The usual approach has been that any technology that produces the best results at the experimental level is superior, and that is what should be offered to the farmer. Thus, technology was generated and efforts made to transfer it without proper knowledge of small farmers.

Failure to consider the actual circumstances under which small farms operate has seriously affected the working methods of the services responsible for improving those farmers' level of technology. Examination of extension publications and training activities shows that an assumption that strongly influenced the operation of extension services is the idea that the main factor hindering the adoption of technology is the farmer's lack of information regarding appropriate techniques for improving the operation of his production system. This may indeed be the case in some situations and the provision of information may be useful in producing change. In other cases, however, the lack of information on technology may not be the factor holding back the improvement of the production system.

The idea that the supply of technology should be well suited to a demand, carries with it the idea that the technology to be transferred must be workable and that those responsible for transferring it must have the necessary technical skills. It is thus obvious that the extensionists must be thoroughly familiar with the practical aspects of small-farm management and the production systems used by small farmers. (Pineiro et al. 1981).

A Strategy for Dealing with Multiple Technological Demands

An analysis of the causes of the lack of coordination between the generation of technology and its effective adoption at the small-farm level brings to light another factor that has negatively influenced organization and operation of technology generation and transfer effort. The assumption was that the demand for technology was uniform enough that a single institutional organization and a single operating strategy, with few variations, would adequately cover different requirements. Thus, such an institution could, by itself, produce changes among different types of producers, different regions and even different countries. It can be shown, however, that failure to adapt to different cultural patterns, financial resources and agroproduction situations made it difficult to bring about significant changes in terms of the adoption of technologies on the part of small farmers.

What is needed is a technology generation and transfer organization and a methodology that will make it possible to recognize and classify the different types of small farmers concerned. Then, and only then, can such an organizational design generate and make available to farmers an appropriate technology which they will then be able to adapt.

The Contribution of Systems Theory to Agricultural Research and Rural Development

I have already pointed out the need to understand the complexity and diversity of small production units. A look at the methodology used to generate and transfer technology shows that it was based on another assumption -- that the production unit could be operated according to a single criterion or to simple decision-making criteria pertaining to the production systems. This approach strongly influenced the organization, design and execution of traditional agricultural research. Those responsible thought that significant changes in the small-farm production system could be produced through action at the level of only some of its elements. This may explain, for example, the discrepancy between the availability of improved crop varieties and their actual use by small farmers.

This one-dimensional approach on the part of traditional agricultural research influenced the methods used for transfering technology. Specifically, there was a tendency to oversimplify the analysis of the factors leading to technology adoption. One effect of this oversimplification was that the supply of technology was based on the assumption that decision-making was based mainly on a desire to maximize production efficiency. Consequently, the idea of increasing physical yields was the main argument used by extension personnel in their work with small farmers.

This short-sighted approach on the part of researchers and extensionists was abandoned because the results obtained from its application were not satisfactory and did not respond to the complex issues raised. Eventually, a multidimensional analytical model was developed to provide a systems approach to agricultural research and rural development.

A process of change, in which interesting ideas and proposals have been put forth, is now underway, but much remains to be done. On-farm research is making a significant contribution towards reducing the isolation of many agricultural research centres. Such research provides valid proposals for solving small farmers' problems by taking into account the circumstances in which they are operating. Further work must be done, however, in the search for a better fit between activities carried out under this approach and activities relating to the generation of basic technological know-how, and between on-farm research and all the other activities involved in the transfer of technology.

In the same way, the system-wide approach to the problems of rural development led to considerable progress. Rural development problems were dealt with in an integrated fashion, following the strategy of integrated rural development, in which priority is given to the man-family-land relationship, rather than only to aspects related to production. Integrated rural development programmes offer a valid alternative for finding solutions that allow the small farmer and his family to improve their standard of living.

Without losing sight of the desired goals, it is essential that designers of rural development projects give careful consideration to the political decisions involved and the financial resources available, because those elements will determine the future of such projects.

CONSIDERATIONS AND SUGGESTIONS FOR THE FUTURE

Technology for the Small-Farm System

Because of the complexity, diversity and instability of the small-farm system, great care is required in selecting the external stimuli to be used to produce changes in performance. Sufficient resources must be assigned (and the necessary decisions taken to make it possible) to start low-input technology generation and transfer programmes, which are suited to the circumstances of the different types of small farms and which can improve their productivity incomes while conserving natural resources (Renard et al. 1983).

Technology, of course, must be considered within a broader frame of reference, recognizing that the processes leading to adoption of technology take place within processes of social and economic change that are broader than those involved in the mere modification of a technological pattern (Pineiro et al. 1981).

Thus, it is important to recognize that the adoption of technology calls for the convergence, at the small-farm level, not only of adequate technology, but also of markets, prices, availability of inputs, adequate credit and everything else that fuels the operation and growth of the production system.

In brief, the issue is that technology, in and of itself, is not sufficient to bring about significant changes in the current economic and social situation of small farmers in Latin America. It can, however, be the central component of a start-up strategy designed to strengthen small farms, favouring their differentiation (Figure 2.1) towards a situation in which they will be able to adopt other technology and in which they will have greater management capability -- all of which would foster savings, self-sustained growth and improved living standards.

The Integration of Research and Technology Transfer

The organization of traditional agricultural research systems is such that they do things their own way, separately and differently from the technology transfer systems. Because there is no linkage between the two, there has been an unsatisfactory flow of workable technology between research centres and production units. This is partly due to the fact that extensionists have not participated at the technology generation stage. At the same time, researchers have neither been responsible for, nor participated in, the transfer-adoption stage.

Generation and transfer of technology must be part of a single process. The differences between the stages in this process consist only of allocation of resources. In some stages, resources are mainly devoted to activities relating to the generation and adaptation of technological know-how designed to improve the ability of farmers to combine their resources and manage their production systems. In other stages, resources are mainly devoted to activities that have more to do with transferring that know-how and making it available at the farm level. In other words, the linkage between research and extension is not so much because the two are in the same institution and under the same chain of command, as because they share a common objective and follow a common procedure in achieving it. This approach is much broader than the traditional one. Research is now given the responsibility of contributing towards improving the technological level of farmers.

In this context, on-farm research programmes offer an attractive alternative. Traditional research and extension programmes, as I emphasized earlier, evolved from the concept of extension acting as a link between research and the farmer. What is needed is to make extension personnel active participants in the research. This option may be accomplished through merger of research and extension components into a research and development organization.

This type of organizational arrangement will provide a better fit between research efforts and the technological needs of farmers through closer contact between those responsible for technology generation and the technology users (farmers and technical personnel working directly with them).

Organizational rearrangements, however will not be enough by themselves; there is also the need to adjust institutional objectives, mandates and internal policies.

Farmers' Organizations as Key Tools

Bringing research and technology transfer to the farm level, however appropriate from a strategic point of view, poses a significant problem for Third World institutions. The approach is much more resource-intensive than the traditional approach, putting increased demand on budgetary and management resources, both of which may be in short supply in a developing country.

A closer functional link between research and technology transfer systems and farmer organizations constitutes an important alternative to explore. Such a linkage will not only facilitate the implementation of decentralized research and speed up information flow, but also provide a platform for an organized demand for technological services and eventually constitute -- directly or indirectly, as a lobbying power -- a source of funding for needed research. A closer and more formal link between technological institutions and farmer organizations will also enhance coherence between technology and other rural development strategies, such as marketing, health and education services, and infrastructural development.

Better Public-Private Sector Links as a Stimulus

In most Latin American countries, agricultural research and the transfer of its findings have been done by public-sector institutions. However, partly because of limited results produced by this sector, particularly as a result of what may be called the region-wide restriction of resources for agricultural research (Renard et al. 1983), and partly because of the particular characteristics of recent technological development, the private sector is playing a growing role in the development, dissemination and utilization of improved technology for agricultural production.

The growing participation of the private sector in organizing and implementing technology transfer brings to light the need for a coordinated, non-competitive relationship between the public and the private sectors. In addition to opening the way for broader technological coverage of farmers having different capacities for acquiring technology (Indarte 1982), the establishment of a proper articulation between the two sectors would enable the public sector to free some of its resources which could then be concentrated on serving the smaller farmers.

Biotechnology's Impact: New Options for the Small Farmer

Finally, I would like to comment on a revolution -- in terms of the scope and prospects for technological innovation on small farms -- that is being brought about by the possibility of manipulating living organisms, following procedures that fall within the research field of biotechnology.

The concept of direct biological manipulation is not a new one. The ideas of improving animal and plant species through selection and crossbreeding, taking advantage of the genetic variety of living organisms; of chemical recombination, using the capacity of some micro-organisms to produce fermentation under given environmental conditions; and of utilizing bacteria that have a capacity to fix nitrogen in order to improve the production of legumes, have long been used in agricultural research.

The great change represented by today's biotechnology lies in the fact that, contrary to the traditional biotechnology in which improvement was based on selection, greater knowledge of the molecular structure of living organisms allows for direct manipulation and modification of cell structure. In other words, biotechnology treats living organisms like machines and assumes that if one can understand their composition, one can modify and redesign them in order to make them serve specific purposes (Janvry et al. 1987).

Although it may seem that there is too great a distance between the advances in biotechnology and the technological needs of small farmers, there are already real possibilities. Cell and tissue cultures, for example, make it possible to multiply high-yield, disease-free local planting material, thus reducing the cost involved in using agrochemical products. Genetic engineering makes it possible to introduce genes for DNA or to manipulate plant tolerance for salinity. These two examples alone are sufficient to give an idea of the potentially tremendous impact that biotechnology can have on small farm systems. In the first case, there is the possibility of improving farm family nutrition by increasing the protein content of traditional foods. In the second case, there is the possibility of farming lands now unproductive because of their high salt content, or of improving the productivity of lands that are undergoing a salinization process because of poor management.

Biotechnology presents a challenge to agricultural research in Latin America, and the progress made in this area in more developed countries should draw attention to the need for national research systems to realize the importance of adopting, adapting, and developing biotechnological products, processes and inputs that can be useful to farmers. Uneven development of biotechnology in countries that are at different stages of development will further widen the technology gap between them, thus increasing the vulnerability of the less-developed countries and making them more dependent on external technology.

By the same token, it is also important to recognize the potentially high risk inherent in biotechnology: it can be more fully exploited by larger-scale farmers. Because of its nature, moreover, biotechnology is highly susceptible to being appropriated by commercial firms in the private sector, thus benefitting those producers with the greatest purchasing power. National institutions, particularly public agricultural research institutions, should monitor advances in biotechnology and protect society from any risks of inequity, without discouraging the investment required to sustain efforts in that direction (IICA 1987).

Table 2.1

FAMILY AND HIRED LABOUR PERCENTAGES FOR THE SMALL-FARM SECTOR AND ALL OTHER AGRICULTURE IN SELECTED LATIN AMERICAN COUNTRIES

	Small farmers (%)	All other agriculture (%)	Total (%)
Brazil (1970)			
Family	92.6	62.9	85.0
Hired	7.4	37.1	15.0
Ecuador			
Family	76.2	39.0	66.1
Hired	23.8	61.0	33.9
Mexico			
Family	72.7	47.1	67.7
Hired	27.3	52.9	32.3
Panama			
Family	79.8	41.5	65.1
Hired	20.2	58.5	34.9

Source: CEPAL-FAO 1986.

Table 2.2

STRUCTURE OF FAMILY INCOME
(PERCENTAGE OF INCOME) BY FARM SIZE

Chile (1974 -1975)	Less than 1 ha	5 to 10 ha
Agricultural activities on the farm:	27.0	79.0
Non-agricultural activities on the farm:	3.0	7.0
Non-farm activities:	53.0	13.0
Others:	17.0	2.0
Guatemala 1 (1978)	**Less than 1 ha**	**1 to 10 ha**
Agricultural activities on the farm:	19.0	45.0
Non-agricultural activities on the farm:	12.0	4.0
Non-farm activities:	65.0	46.0
Others:	4.0	5.0
Guatemala 2 (1979)	**Less than 1 ha**	**1 to 10 ha**
Agricultural activities on the farm:	18.0	40.0
Non-agricultural activities on the farm:	6.0	5.0
Non-farm activities:	74.0	47.0
Others:	2.0	5.0
Peru (1974)	**Less than 3.5 ha**	**3.5 to 11 ha**
Agricultural activities on the farm:	15.8	46.7
Non-agricultural activities on the farm:	22.0	23.0
Non-farm activities:	58.0	29.0
Others:	5.0	5.0

Source: Selected case studies (CEPAL-FAO 1986).

Table 2.3

SHARE OF SMALL FARMS IN THE PRODUCTION (% OF PRODUCTION) OF COFFEE, COCOA AND TOTAL AGRICULTURAL PRODUCTION IN SELECTED LATIN AMERICAN COUNTRIES

Country	Coffee		Cocoa		Total agricultural production	
	Year	(%)	Year	(%)	Year	(%)
Brazil	1980	40.3	1980	32.8	1980	39.6
Colombia	1970	29.5		n.a	1981	44.1
Mexico	1970	53.8	1970	45.9	1970	46.9
El Salvador	1971	19.9				
Peru	1972	54.8	1972	67.5	1977	54.9
Venezuela	1976	63.2	1976	69.1		
Bolivia		n.a		n.a	1977	80.0
Chile					1980	37.8

Source: CEPAL-FAO 1986.

Table 2.4

SHARE OF SMALL FARMS IN THE PRODUCTION OF SELECTED FOOD CROPS IN SIX LATIN AMERICAN COUNTRIES (% OF PRODUCTION AND TOTAL LAND)

Crops	Brazil (1980)		Colombia (1970)	
	Production	Surface	Production	Surface
Maize	52.0	53.0	47.0	85.0
Beans	63.0	61.0	69.0	81.0
Potatoes	-	-	25.0	67.0
Lentils	-	-	-	-
Rice	23.0	25.0	13.0	34.0
Barley	-	-	20.0	30.0
Cassava	68.0	64.0	60.0	80.0
Plantains	-	-	85.0	90.0

Source: CEPAL-FAO 1986, Table 4, p.23

(Continued)

Table 2.4 (Cont.)

SHARE OF SMALL FARMS IN THE PRODUCTION OF SELECTED FOOD CROPS IN SIX LATIN AMERICAN COUNTRIES (% OF PRODUCTION AND TOTAL LAND)

Crops	Costa Rica (1973)		Chile (1980)	
	Production	Surface	Production	Surface
Maize	60.0	52.0	44.0	51.0
Beans	54.0	54.0	77.0	75.0
Potatoes	55.0	57.0	73.0	83.0
Lentils	-	-	47.0	47.0
Rice	10.0	24.0	55.0	64.0
Barley	-	-	-	-
Cassava	-	-	-	-
Plantains	-	-	-	-

Source: CEPAL-FAO 1986, Table 4, p.23

(Continued)

Table 2.4 (Cont.)

SHARE OF SMALL FARMS IN THE PRODUCTION OF SELECTED FOOD CROPS IN SIX LATIN AMERICAN COUNTRIES (% OF PRODUCTION AND TOTAL LAND)

Crops	Ecuador (1974)		Panama (1984)	
	Production	Surface	Production	Surface
Maize	45.0	54.0	80.0	91.0
Beans	61.0	67.0	-	-
Potatoes	48.0	49.0	-	-
Lentils	37.0	44.0	-	-
Rice	34.0	34.0	37.0	67.0
Barley	54.0	56.0	-	-
Cassava	33.0	36.0	-	-
Plantains	17.0	17.0	-	-

Source: CEPAL-FAO 1986, Table 4, p.23

Figure 2.1

THE PROCESS OF DIFFERENTIATION RESULTING FROM THE EFFECTS OF EXTERNAL FACTORS ON SMALL FARMS

Breakdown of the System

Loss of control over resources of production, transformation of the farmer into a semi-proletarian, squatter or proletarian, squatter or proletarian with no direct access to land.

Precariously Balanced System

Production based on family labor and use of rudimentary means of production. Production is mainly devoted to consumption on the farm itself. Sporadic availability of surpluses. When surpluses do exist, priority is given to meeting needs for food, clothing, housing, and continuation of the production process.

Strengthening the System

Improvement of the level of production, capitalization, possibility of growth through investment of surpluses, high percentage of production directed to market.

BIBLIOGRAPHY

CEPAL, (Economic Commission for Latin America and the Caribbean)-FAO, Agricultura campesina en America Latina y el Caribe, 1986.

FAO, El minifundo en America Latina, Santiago, Chile, 1987.

IICA, Innovacion tecnologica y desarrollo agropecuario en America Latina y el Caribe: Desafios y oportunida- des, Documento No. 6, San Jose, Costa Rica, 1987.

Indarte, E., Articulacion entre generacion y transferencia de tecnologia agropecuaria, Internal IICA document, San Jose, Costa Rica, 1982.

Janvry, A. de., D. Runsten and E. Sadoulet. Technological Innovations in Latin American Agriculture, IICA Program Papers, Series No. 4, San Jose, Costa Rica, 1987.

Maunder, A.H., La extension agricola, FAO, Rome, 1973.

Pineiro, M., J. Chapman, and E. Trigo, Tema sobre el desarrollo de tecnologias para pequenos productores campesinos, Desarrollo Rural en las Americas. 1981.

Murmis, M. and G. Cucullu, Tipologia de pequenos productores campe-sinos en America Latina, IICA, PROTAAL, Document no. 55, San Jose, Costa Rica, 1980.

Renard, A., J. Weber, S. Muani and E. Indarte, Extension agricola, Universidad Nacional del Rosario, Argentina, 1983.

Ruttan, V.W., Agricultural Research Policy and Development, Research and Technology Paper No. 2, FAO, Rome, 1987.

Trigo, E., M. Pineiro and J. Ardila, Organizacion de la investigacion agropecuaria en America Latina, IICA Research and Development Series No. 2, San Jose, Costa Rica, 1982.

CHAPTER THREE

ORGANIZATION AND MANAGEMENT OF RESEARCH FOR RESOURCE-POOR FARMERS

Alexander von der Osten, Peter Ewell and Deborah Merrill-Sands 1/

National agricultural research systems (NARS) in the Third World are expected to produce technology appropriate for a wide range of clients. Table 3.1, adapted from a paper by Chambers (1988), summarizes the agroclimatic and socio-economic characteristics of three general types of agriculture.

The first type consists of the high-input, high-yielding production systems found primarily in the industrialized countries, but also in specialized Third World enclaves. These farmers face problems of overproduction, mounting surpluses and escalating costs.

The second type is found in high-capacity areas of the tropics, particularly where irrigation is available. Applied scientific research has had dramatic impact on the yields of major food grains. Large and small farmers, as well as urban consumers, have benefitted from the development of agricultural technologies, popularly known as the green revolution, but at a cost of increased dependence on purchased inputs.

1/ Executive Secretary of the Consultative Group on International Agricultural Research, Washington, D.C., U.S.A.; Regional Social Scientist for Africa at the International Potato Center, Nairobi, Kenya; Regional Officer of the International Service for National Agricultural Research, The Hague, Netherlands.

The poorest and most vulnerable segment of the world's population falls into the third category -- rural households with few resources beyond the labour of their own families. They work in areas with low and uncertain rainfall, little irrigation, steep slopes, poor roads, and many other limitations. Yields are low and uncertain, and the land, forest, and other resources are rapidly being degraded. Nevertheless, their diverse production systems are tributes to human ingenuity, and there is much untapped potential. This seminar underscores the particular need to increase food production in these marginal areas.

Agricultural research faces important challenges to provide technology to benefit all three groups. Most research systems have been organized and managed to provide technology to increase the productivity of the first two types of agriculture. Few institutions have focussed effectively on the needs of small, resource-poor farmers. This failure stems in part from the inherent difficulty of the task - it is not easy to develop sustainable technology for the heterogeneous agro-ecological and socio-economic conditions of such farmers. But it also stems from institutional constraints: resource-poor farmers have not had a voice - a means to express their demands for technology - within the research system, and, as a result, their needs have not been effectively addressed.

Over the past 15 years, many national research institutes -- often in cooperation with international donors and research centres -- have tried to respond more effectively to the needs of reource-poor farmers by building up their capacity for on-farm adaptive research designed to link farmers directly with the research process.

These on-farm research programmes have faced complex problems of organization and management. Interdisciplinary teams must be sent to the field, far from established centres, to do surveys, experiments, and other activities. Special skills and training are required, as well as reliable systems of logistical support. On-farm work must be programmed and supervised to ensure that it complements, rather than duplicates, research on experiment stations. Effective ways to carry the results to the farmers through extension departments and development projects must be developed. These institutional issues become increasingly important as on-farm research is incorporated into regular programmes of the NARs, as the first flush of enthusiasm for the new programmes encounters deep-seated habits and procedures, and as support from donors for special pilot projects is reduced. Chronic problems of implementation have led some scientists and managers to question the long-term sustainability of the farming systems approach.

To date, on-farm research has emphasized methodologies to develop technologies appropriate to farmers' conditions. Less attention has been given to the problems of institutionalization, namely, how to build sustainable programmes appropriate to the context of the larger research systems of which they are a part. A close analogy can be drawn between the process of adapting a new technology to small farmers' conditions and the process of establishing a new approach within a research institution.

It is widely accepted that no single improved crop variety, no matter how well it does on an experiment station, will necessarily be better than the farmers' existing varieties when grown in their fields. Similarly, different types of on-farm research programmes must be developed for different institutional settings. We can take the analogy one step further. A technology heavily dependent on inputs from outside the immediate region, and sensitive to variations in the environment, will not be sustainable on small farms. Moreover, a farming systems programme heavily dependent on foreign personnel, funds, and methodologies, and which cannot withstand problems such as fluctuations in the operating budget, will not be sustainable without modification.

There can be no simple formula, no blueprint, for successful on-farm research. Nevertheless, considerable experience is now available for research managers to draw upon (Merrill-Sands 1987).

ISNAR'S STUDY OF ON-FARM, CLIENT-ORIENTED RESEARCH (OFCOR)

Since 1986, ISNAR has done major research designed to draw practical lessons from the experiences of NARS in organizing and managing on-farm client-oriented research (OFCOR). OFCOR is a research approach that uses on-farm trials, formal and informal surveys, and meetings with farmers in order to diagnose problems as the farm level, rank them, and then design and test appropriate technologies for solving them. OFCOR foucsses on the needs of resource-poor farmers and is committed to bringing them more fully into the research process.

The ISNAR research was initiated in response to requests from national research managers interested in practical advice on the problems of implementation of OFCOR programmes. The research was a collaborative effort on nine NARS and ISNAR projects, with the Government of Italy and the Rockefeller Foundation providing support.

Because the project focusses on how to implement and sustain OFCOR, national systems where on-farm research had been successfully established for several years were chosen as case studies. Senior professionals in each NARS interviewed key people involved with on-farm research and collected a range of other data, following guidelines developed at ISNAR. A special series of ISNAR publications will comprise the nine case studies, plus comparative studies by researchers at ISNAR. A list of the case studies is in Appendix A.

FUNCTIONS OF OFCOR WITHIN A LARGER RESEARCH SYSTEM

The OFCOR programmes are designed to carry out various roles, or "functions," within the context of a larger research system. Each institute has its own priorities. Most have set up on-farm research programmes, which they hope will perform some combination of the following seven functions (Merrill-Sands 1987):

1. Support, within research, a problem-solving approach, which is fundamentally oriented to farmers as the primary clients of research.
2. Contribute to the application of an inter-disciplinary systems perspective.
3. Characterize major farming systems and client groups, using agro-ecological and socio-economic criteria, in order to diagnose priority problems of production. This should lead to the identification of key opportunities for research with the objective of improving the productivity and/or stability, of those systems.
4. Adapt existing technologies or contribute to the development of alternative technologies for farmers sharing common production problems, by conducting experiments in farmers' fields.
5. Promote the participation of farmers in research as collaborators, testers and evaluators of alternative technologies.
6. Provide feedback into priority-setting, planning and programming of research. The goal is to integrate research on experimental stations and on farms into a coherent programme focussed on farmers' needs.
7. Promote collaboration with extension and development agencies to improve the efficiency of generating and diffusing technology.

The relative emphasis placed on each function varies not only between countries, but also among administrators and scientists, and among scientists in different programmes. The rest of this paper will focus on those functions that depend particularly on clear organization and effective management within the research system.

BUILDING AN INTERDISCIPLINARY PERSPECTIVE

The mandate to adapt research to the agro-ecological and socio-economic conditions of small farmers involves a range of research activities broader than the conventional limits of any single discipline. Innovations within complex farming systems require cooperation between specialists in crop and animal production, as well as in soil and water management and other disciplines. Perspectives and methods from the social sciences can identify priority problems within the farmers' context, and help to reorient priorities as conditions change. Large interdisciplinary teams of natural scientists, social scientists, technicians, and field assistants, as well as extension and other specialists, are not easy to manage effectively. Even if appropriate specialists and sufficient resources are available -- and they seldom are -- posting people from different disciplines to the same OFCOR programme does not guarantee that they will work together productively. Active, creative research management is required.

Strategies for Achieving Disciplinary Breadth

The programmes reviewed above have developed different strategies for achieving disciplinary breadth (Ewell 1988). At one extreme, one or two agronomists with special training are responsible for on-farm research in an entire province. At the other, large interdisciplinary teams with as many as 15 or 20 members are responsible for a broad range of projects. The approaches can be categorized into three broad types:

1. Reliance on Researchers From a Single Technical Discipline. In nearly half of the 21 separate programmes in the nine countries studied, most of the field research was carried out by generalists. They are usually agronomists, or animal scientists in livestock projects, who have received additional training in surveys and in simple techniques of economic evaluation. They have often been supported by specialists from an experiment station or national office. Social scientists have participated, but only in certain specialized activities. Many are concentrated at early stages of the on-farm research process in the planning, design analysis of diagnostic surveys, and in the development of methodology. This work is intended to focus subsequent experimental work on the solution of the most pressing problems which farmers face.

2. The Minimal Pair of Disciplines. Agronomists have been teamed with agricultural economists in the field in about 15% of the programmes studied. They do surveys and other diagnostic studies and then systematically screen component technologies according to agronomic and economic criteria.
3. Farming Systems Teams with a Broader Range of Disciplines. One-third of the programmes have mounted broader farming system teams, which include specialists from additional disciplines, such as animal science, forestry, nutrition and anthropology.

Sustaining an Interdisciplinary Focus

Guidelines for on-farm research stress interdisciplinary cooperation. The case studies show that unless the participation of different disciplines is explicitly written into job descriptions and work plans, it is not easy to sustain institutional support for this perspective. The history of the Production Research Program (PIP) in Ecuador illustrates the issues involved.

PIP was established in 1977 within the national agricultural research institute to complement on-farm research and regional trials. It was organized in close cooperation with CIMMYT and received support from other international centres and universities. A central group of agricultural economists developed a methodology based on a sequence of informal and formal surveys and on-farm experiments. It was implemented by teams of one or two field agronomists, who received intensive training. The headquarters staff negotiated personally with key scientists from the commodity programmes at the regional experiment stations to coordinate support for particular lines of research.

This centralized management approach successfully launched a highly visible pilot programme, which was extended to 10 regions of Ecuador within a few years. The problem has been to maintain dynamism after the original group of founders moved on to other jobs, as international support was reduced, and as the small OFCOR teams of generalists were institutionalized into normal programmes of the regional experiment stations.

Ten years after its founding, no social scientists were involved with the programme, and the direct participation of commodity scientists in on-farm research had been reduced significantly. Training programmes had fallen behind staff turnover and by 1986 only 20% of the field agronomists had participated in CIMMYT's basic course on the methodology. Valuable on-farm research continues to be done, but its scope has narrowed considerably. Few or no surveys were done after the initial phase of diagnosis, and field research methods have fallen into routine patterns. The PIP programme was under pressure from the commodity scientists on the stations to concentrate on the routine screening of their technology. Several reviews of the programme have strongly recommended it should be reorganized in accord with its original goals.

Broadening the Agenda of Social Scientists

Several of the programmes in the case studies have attempted to extend the participation of social scientists beyond certain stages of the research process. It is widely accepted that economists have a role early in a project for planning, diagnostic studies, and the development of methodology, and perhaps again later for evaluations. It has been more difficult to institutionalize a continuing role for a broader range of disciplines in the field research process. The experience of the OFCOR programme in Zambia illustrates ways in which this can be done. Each of seven provincial Adaptive Research Planning Teams (ARPT) includes at least one economist. Sociologists and nutritionists are located in separate teams with responsibility for several provinces. They carry out special studies and cooperate with the teams on methodological issues as needed. These have included the improvement of procedures used to select farmers, and the encouragement of the more active participation of farmers in the identification of priority problems and in the evaluation of results.

The maintenance of a broad research agenda and the systematic incorporation of information about the farm households' conditions and their demands for new technology has proved difficult to maintain. We can conclude that a broad OFCOR agenda must have the strong support of senior management, sustained scientific leadership, and continuous training of the field staff to prevent problems of this kind.

Fostering Regular Involvement of Senior Scientists

Most OFCOR programmes try to supplement the expertise of their field staff with the regular support of senior scientists. An interesting mechanism for doing so was developed in Nepal where several OFCOR programmes date back to the early 1970s. The scientists responsible for on-farm research are based at experiment stations. The supervision and support of field technicians at farming-systems research sites in the hill regions is a daunting logistical problem. Regular field visits, called

Group Treks, were pioneered by the Lumle and Pakhribas Agricultural Centres -- regional research and extension centres funded by the British Overseas Development Administration. The treks have since been adopted by the national research system. An interdisciplinary tour of the farming systems sites is organized before the beginning of each agricultural cycle. Senior scientists, technicians, and assistants travel together. At each site they interview farmers, women, local officials and other key informants. Then they meet together to develop, on the spot, the research plan for the next period.

Other countries have developed methods to get interdisciplinary teams of senior scientists into the field. Most, however, are used to setting primarily initial priorities for research in a region. The Group Treks are unique as a managemnt tool because they are used in regular, year-to-year programming. Nevertheless, there have been difficulties maintaining the dynamism of the events. One is that scientists from various disciplines need to be convinced that it is worth their time to go repeatedly into the field and to be closely involved in the process of adapting technology to a particular region. Another difficulty is that the regular programmes of the Nepalese government must abide by regular civil service rules and do not have the same flexible control over their operating budgets as do the programmes funded by foreign donors. The per diem allowances do not usually cover the expenses of a Group Trek. Some scientists are willing to sacrifice for a special occasion of particular interest, but not on a regular basis.

Lessons Learned:

Several lessons for building an interdisciplinary perspective can be drawn from the experiences in the nine countries:

Maintain focus on the clients: Research managers must build special events and meetings into the regular planning and programming process to maintain focus on the clients. If left to the initiative of individual scientists, less emphasis is placed on the specific problems of the complex farming systems of small farmers. There is a strong tendency for natural scientists to become increasingly specialized as they move up the career ladder. An explicit regional approach to priority-setting and planning in many OFCOR programmes has proved useful in this regard.

Support from senior scientists: Methodological innovations are difficult to sustain without the active ongoing participation of senior scientists from various disciplines. Consistent support is needed at the policy level (so that scientists derive professional benefits from on-farm research), and on the level of implementation (so that the necessary resources and logistical support are made available).

OFCOR programmes often depend heavily on junior staff members and on generalists for day-to-day field research. Management must support these people so that they are not cast in the role of mere low-status, testers of technology. This is important so that the knowledge they gain from their direct contact with farmers can be used in the programming and evaluation of the broader research programme.

Sustain social scientists in field research: In general, social scientists are a scarce resource in most NARS. Unless their ongoing participation in on-farm research gets high priority, able senior professionals tend to get drawn off into planning departments, the preparation of proposals to foreign donors, and other functions. There is strong evidence that social scientists should participate as regular, ongoing members of OFCOR programmes. They help to maintain focus on the dynamic problems of small farmers and to prevent OFCOR from slipping into routine on-farm testing.

Need for training: Continuous training must have high priority, both to broaden the skills of the field staff to do interdisciplinary research and to train new scientists and technicians as they join the programme.

INTEGRATION OF OFCOR WITH EXPERIMENT STATION RESEARCH

On-farm research should complement the work done on experiment stations because an OFCOR programme cannot survive as a stable component of a research system if it operates in isolation. In theory, researchers in each programme should provide their colleagues with information and services vital to the achievement of their common goals: the generation and transfer of technology appropriate to the needs of small farmers with limited resources. This may seem self-evident, but many programmes have not been able to maintain productive interaction between scientists working on farms and those on experiment stations.

On-farm researchers should carry out several functions useful to scientists on the stations. They can screen, test, and evaluate promising component technologies in farmers' fields. They can carry the process a step further with adaptive research, developing new variants on technologies in cooperation with farmers. Finally, OFCOR can provide feedback -- information from surveys, experiments, and what they learn directly from the farmers -- which can lead to more relevant priorities for on-station research.

At the same time, scientists in the commodity programmes and departments should provide critical support to OFCOR. They have the facilities and specialized skills to do applied research and to generate component technologies appropriate to specified conditions. They can also support their colleagues in the field with various services, such as diagnosising constraints, designing experiments, and analysising data, etc.

Common Areas of Conflict

The nine case studies documented many sources of friction between OFCOR scientists and their colleagues in on-station research. In more than half of the programmes studied, on-station researchers perceived OFCOR as competing with their own work; only three had a shared consensus of what their mutual roles should be. Differences in goals and attitudes lead to disputes over resources, over priorities in the planning and programming of research, and over the legitimacy and interpretation of results. The resolution of these problems is critical if OFCOR is to achieve long-term institutional stability. The comparative study of the cases identified a number of common problems, as well as mechanisms that have helped managers to build more effective linkages (Merrill-Sands and McAllister 1988).

Conflicts Over the Research Agenda

Some of the problems reflect basic differences in goals and scientific attitudes. Conventional agricultural research focusses on individual components of a production system and seeks to design the best technology, usually defined as that which has the highest yields under broadly specified constraints. OFCOR orients research towards farmers, and seeks to develop better technology which is suitable and sustainable under their circumstances. On-farm research has usually been created as an explicit response to the fact that on-station research had not succeeded in developing technology appropriate to resource-poor farmers' needs.

In this context, it has often proved easier to set up separate programmes of applied and adaptive research than to agree on how to integrate them into a single process. When challenged, on-station researchers defend traditional, standard criteria of what constitutes good science. They criticize research under the less-controlled conditions on farms because of high loss rates and high coefficients of variation. They are unsure about how to interpret socio-economic surveys and other types of data with which they are unfamiliar. They are reluctant to modify their own research agenda, or to adjust the standards used to approve technology for release. For their part, OFCOR researchers protect their own agenda and resist being drawn into multi-locational testing and other programmes directed from the stations.

Conflicts Over Status and Legitimacy

Research agenda issues are reinforced by conflicts over the status and legitimacy of the scientists involved. It is usually young, relatively inexperienced researchers who are willing to be posted in isolated locations to do on-farm research. It is difficult for them to speak with authority to their senior colleagues at the stations. These differences in status are accentuated by better opportunities for short courses and advanced degree training available from within station-based programmes. The career ladder usually leads into the regular programmes and departments, siphoning experienced people away from OFCOR. There are exceptions, particularly in Africa, where it is the on-farm researchers who have been relatively privileged. It should be mentioned that the OFCOR programmes studied by ISNAR have been funded by international donors, who have provided vehicles, travel and housing allowances, overseas conferences, and other resources not always available on the stations.

Barriers to Cooperation

Close cooperation between on-farm and on-station research benefits an institution in the long run, but there are immediate costs to the scientists themselves. Such cooperation distracts them from their research projects and involves extra work, travel and meetings. These costs are often not fully compensated by the institutions; as a result, two-thirds of the cases reported that scarcity of funds and personnel limited opportunities for cooperation. Most professional awards go to achievements in specialized research so that time invested in collaborative work is less likely to be recognized.

Active, innovative management can develop ways to overcome many of these problems, or at least minimize their effects. This is a critical area. OFCOR programmes can only succeed in meeting the needs of their clients if they are successfully integrated within their larger research systems.

Types of Organizational Arrangements

The type and frequency of contact between on-farm and on-station researchers varies depending on the position of an OFCOR programme within the larger research institution. There is often a trade-off between staying in close contact with farmers and maintaining close linkages with other scientists. Four major types of arrangements were found in the case studies, each of which requires a different strategy for maintaining a reasonable balance.

On-Farm Teams Linked to Regional Experiment Stations

The most common arrangement in the case studies is the posting of OFCOR scientists to a number of field teams, each linked to a regional experiment station. In Guatemala, the planning and programming of research is totally regionalized. In Ecuador, Zambia and Senegal, a national OFCOR office maintains some control over the research agenda as well as the methodology.

Putting on-farm research in a special unit with its own professional and administrative identity facilitates the development of a programme oriented to farmers' needs. Nevertheless, active, creative management is needed to maintain effective two-way communication with on-station researchers. In Ecuador, the commodity scientists have attempted to capture the on-farm programme for their own screening purposes. In Guatemala, a clearly defined research programme and regular planning and review meetings failed to eliminate conflicts over status and legitimacy.

OFCOR Programmes Directed Out of Stations

In Bangladesh, Nepal and Zimbabwe, OFCOR field research is done in designated areas. The scientists who plan and implement this work are based at regional or national experiment stations. Formal and informal contacts with their colleagues in the national programmes and departments are relatively easy to arrange, but still require active encouragement. However, special attention and certain mechanisms, such as Nepal's Group Trek, are needed to keep them in direct contact with farmers.

Scientists Responsible for Both On-Station and On-Farm Research

When the same scientists are responsible for both on-station and on-farm research, linkage is obviously not an issue. Such cases, which include programmes in Indonesia, Nepal, Zimbabwe and Panama, have faced two different kinds of problems. The first has been the organization and subsequent logistical support of many separate research activities. The second problem has been the maintenance of a broad, client-oriented perspective. The OFCOR programmes of the Malang Institute in Indonesia and the Lumle Centre in Nepal held frequent meetings to ensure that all of the participating scientists maintained their focus on common farm-oriented goals.

Special Projects

In Indonesia, Ecuador and Panama, OFCOR programmes have been set up within specially funded projects, often tied to regional development programmes. In these cases, short-term problems of coordination and linkage can be solved directly because research tasks are allocated between experimental plots and farmers' fields according to a single plan. Nevertheless, there is little incentive to develop programmes and institutional arrangements that will continue after the end of the projects.

Mechanisms That Encourage Integration

The case studies demonstrate that the following mechanisms promote integration between on-farm and on-station research. They provide opportunities for interaction, mobilize resources to facilitate the process, and create incentives for the scientists to participate.

Joint Participation in Priority-Setting, Planning, and Review

Senior scientists from research stations have participated in the diagnostic phases of many of the OFCOR programmes. This important first step of setting priorities provides a baseline for developing a complementary research agenda. Nevertheless, it has proved difficult to maintain a spirit of equal and active participation in the year-to-year planning, programming, and review of research, particularly on the stations.

Regular joint meetings and field visits with clear goals, explicit responsibilities, and the authority to implement decisions can prevent each programme from retreating into its own agenda. Such regular events require advance planning and adequate funds from a separate, secure budget, especially in situations where the physical distances between on-station researchers and field-based OFCOR staff is great.

Formal Collaboration On Trials and Surveys

Many conflicts have arisen between on-farm and on-station researchers because they question each other's methods and results. In the research systems studied, some scientists informally collaborated in trials, but without much support from the programmes. Participation in trials in farmers' fields helps on-station researchers appreciate the difficulties as well as the advantages of the approach. Participation in trials on stations enhances the scientific credentials of OFCOR scientists. Collaboration on surveys, in the few cases where it was tried, was particularly valuable because many station-based researchers were unfamiliar with the methods and the interpretation of survey results.

Assignment of Responsibility for Coordination

Effective integration of on-farm and on-station research requires attention to many details that fall outside of routine programme responsibilities. Active coordination is required. Yet coordination is a task with few professional rewards in most research systems. A comparison of the case studies suggests that a two-tiered distribution of explicit responsibility is the best solution. A high-level manager should set the policy and legitimize the goals, and then delegate responsibility for organizing and implementing collaborative activities to the managers responsible for the programming and review of field research.

Formal Guidelines for Allocating Time and Funds

There is a tendency for joint activities to be concentrated in the early phases of an OFCOR project, when enthusiasm is high and external funds are available. Shortages of time and money are often cited as the reasons why the effort is not sustained. Managers can insure that collaborative events continue by including them formally in annual work plans and budgets.

OFCOR AS A LINK BETWEEN RESEARCH, FARMERS AND EXTENSION

The generation and transfer of technology appropriate for poor farmers require links between them and the research programmes. These links provide flows of information from farmers into the research process on the one hand, and the diffusion of information and technology to the farmers, on the other. The OFCOR programmes studied developed a variety of ways to enlist the participation of farmers and also to involve extensionists as farmer representatives. The programmes had problems, however, in developing formal linkages for the diffusion of technology through extension institutions and development programmes.

Participation of Farmers in On-Farm Research

One of the principal goals of on-farm programmes has been to involve farmers directly in the research process. Four distinct modes of participation were identified in the case studies (Biggs 1988). The typology is adapted from one developed by researchers at CIAT (Ashby 1986). Each plays an important role in different types of research.

Contract Participation

In contract participation, farmers are given contracts to provide land for trials, as well as field preparation and (in some cases) other labour. The researchers are interested primarily in representative soils and agroclimatic conditions and attempt to maintain as much control as possible over experimental and non-experimental variables. This type of arrangement is widely used for the testing of new varieties and in soil fertility programmes. The trials often conform to standardized experimental designs, and the active participation of the farmers is discouraged. Researchers do not systematically take advantage of the knowledge and experience of the local people.

Consultative Participation

The most common mode of participation found in the case studies is best described as consultative. OFCOR programmes seek to tailor technology to the agro-ecological and the socio-economic constraints facing different groups of farmers. Surveys and meetings are used to identify problems and needs. Then the scientists design on-farm experiments to test alternative solutions. This consultative relationship is analogous to the relationship between doctors and their patients. On-farm experiments are commonly organized in a sequence of stages from diagnosis to design, testing and verification. The degree of farmers' participation is always determined by the researcher. Social scientists are commonly charged with representing the farmers' point of view in the planning and review process.

The development of research methods appropriate to the consultative mode of participation and their application on a wide scale have been major achievements in all of the countries studied. Nevertheless, in several cases, as procedures became routine, methodologies tended to stagnate. Even in well established programmes, little new information was brought in from farmers after the initial diagnostic stage. Another common issue, somewhat surprisingly, has been the low priority placed on procedures for selecting representative farmers as cooperators. Many researchers accept the suggestions of extension agents, take volunteers at meetings, or turn the problem over to junior field staff without adequate guidance. Samples are easily biased in favour of relatively more prosperous or politically active farmers, which means that the results may not be appropriate for resource-poor farmers who are the intended clients.

Collaborative Participation

In collaborative participation, farmers are regarded as active partners in the research process. They are involved in regular meetings designed to clarify the logic of both their current practices and their demand for new technology. They participate directly in the plannning, execution and evaluation of trials.

The case studies found many examples of close collaborative interaction between researchers and farmers. The best established are livestock projects in Indonesia and Panama and pest management research in Zimbabwe. What these have in common is the continuous, routine monitoring of data on farmers' practices, which has provided opportunities and incentives for close collaboration. In Nepal, and in other countries, local farmers were hired as junior technicians on research teams. In several cases, individual scientists used innovative methods to involve farmers in research. Because such methods are not easily codified into standardized procedures, they are the first things to be cut in a budgetary squeeze. In general, it requires constant attention to incorporate new kinds of data from the farm level into an established research system

Collegiate Participation

Collegiate participation refers to the flow of resources and knowledge in the opposite direction from research-minded farmers, who can be found in every farm community, to research institutions. Although none were documented in the nine case studies, there are examples from all over the world of technology that has been developed by farmers and then spread laterally. Strengthening farmers' capacity for informal research and development can be an effective strategy to complement formal research programmes, particularly where farming systems are complex and diverse (Biggs 1986, Biggs and Clay 1981).

Meetings with Farmers

Most of the OFCOR programmes studied have organized meetings with farmers as part of the research process and as a complement to formal surveys and on-farm experiments. Group tours, such as Group Treks in Nepal and sondeos in Guatemala, get senior researchers into the field to talk with farmers as part of the priority-setting and planning process. Village meetings are used to discuss local problems, to design research programmes and to organize their implementation, and to interpret the applicability of the results. When organized effectively, meetings can be an inexpensive method for gathering valuable information, and feed-back can prevent serious problems from developing in the conduct of on-farm research. Effective management of meetings with farmers requires sensitivity to local values and to the structures of power and influence in a village.

Field days play a role in many kinds of agricultural research programmes. Most of the OFCOR programmes have increased the relevance of field days for small resource-poor farmers by holding them in the fields of the collaborating farmers, who participate directly in the explanations. Some researchers have refined the process further by organizing separate, focussed events. Only farmers and extensionists are invited when the goal is to discuss experimental results and their relevance and priorities for further work. Local political leaders and scientists from the stations are invited to other events designed to publicize OFCOR and to generate support. These meetings are on the border between experiments and demonstrations -- between research and extension.

Participation of Extension in On-Farm Research

Extensionists have participated in OFCOR in two principal roles. The first has been to represent the farmers in order to get information on their conditions and needs incorporated into the research process. Mechanisms have ranged from informal advice on the selection of research zones and farmers to jointly organized programmes. The second role has been for the extensionist to act as a bridge for the formal transfer of information into the larger extension system. Although all of the national research institutes studied expected their on-farm programmes to strengthen the link with extension, this function has been one of the most difficult to fulfill.

Research and extension institutions are not easily married in joint programmes. They often have different operational regions, and different internal structures as regards disciplines, commodity programmes, and so on. In many cases, extension agents are responsible for a broad range of government services in rural areas, of which technology transfer is but one. In several countries, both research and extension have been extensively reorganized in cooperation with foreign donors, without any effective joint planning. A series of more subtle factors -- such as differences in social class, educational level, pay, status, and access to vehicles -- can become serious barriers to effective cooperation.

OFCOR as an Alternative Diffusion Mechanism

A few programmes in the case studies relied on the demonstration effect of a network of on-farm trials to diffuse technology without the formal intervention of extension. For example, in Guatemala, the OFCOR programme worked with the extension department in a few isolated cases, but until recently could not develop a regular working relationship. As an alternative, they incorporated technology transfer into the on-farm research process through demonstrations and field days and relied on the farmers themselves to spread the word. There is evidence that several new crop varieties have spread widely, but that poor farmers in marginal regions have not benefitted. A new externally funded programme is developing formal mechanisms to improve the linkage with the extension department as part of a programme to serve the needs of that client group. This experience is repeated in several other cases.

Informal Cooperation at the Field Level

The field staff members of many OFCOR programmes depend heavily on local extension agents for key functions -- securing the cooperation of local leaders, identifying collaborators, and organizing field days. Friendships and good working relationships have developed on this basis, but there is good evidence that informal cooperation alone is not an effective transfer method. Several provincial OFCOR teams in Ecuador share offices with extension workers but still find it difficult to coordinate activities because operational responsibilities - and even working hours - are different. In Senegal, effective collaboration was developed between a regional OFCOR team and the extension programme of a development agency, but it did not survive the departure of foreign scientists who had independent access to operating funds from a donor.

Participation of OFCOR Staff in Rural Development Projects

Integrated rural development projects usually attempt to achieve a number of parallel goals, of which increased agricultural productivity is only one. Because they do not have research capacity of their own, many have asked OFCOR programmes to cooperate in the development of locally adapted technology. This constitutes a strong institutional demand for OFCOR. We have examples from Ecuador, Senegal, Zambia, Nepal and Indonesia, with a range of results. The advantage of these arrangements is that researchers and extensionists can collaborate closely under a single management structure. Nevertheless, OFCOR field teams can become distracted from their long-term research goals to meet the immediate needs of a project, such as the multiplication of seed or the testing of inputs. Programmes are repeatedly redefined to meet the requests of different donors, and data are not accumulated or interpreted according to consistent criteria.

Participation of Extensionists as Technicians in Research Programmes

The delegation of the routine management of surveys and on-farm trials to local extension agents is tempting, as it offers to increase the geographical coverage of a research programme with existing personnel. Experience in Guatemala and in Zimbabwe suggests, however, that, in addition to their normal duties, extension agents cannot run experimental plots effectively. They don't have the time, the training and experience, the mobility, or the motivation to keep the loss rates and coefficients of variation down. This is a recipe for frustration -- everyone involved ends up feeling that they are wasting time. In Zimbabwe, the managers of the on-farm testing programme learned this lesson and clustered their experiments into a few representative areas, under direct supervision of technicians from the research institute.

A more effective solution is the formal delegation of extension agents to OFCOR programmes as technicians. In Zambia, trials assistants who speak the local languages and who understand local agronomic practices and food preferences have been effective members of on-farm research teams. The original intention was to rotate them and expose as many extension agents as possible to the research process and the new technology. Nevertheless, the Research Branch soon moved to hold onto them, to retain the benefit of their experience and to avoid the costs of constantly retraining new staff members.

Extension Specialists as Scientists in OFCOR Programmes

In Zambia, Research-Extension Liaison Officers are full-fledged members of the provincial OFCOR teams. This arrangement facilitates the flow of information between research and extension. Nevertheless, the position's precise responsibilities have proven difficult to define because the liaision officers have to report to supervisors within both research and extension. This has led to divided loyalties, confusion in job descriptions and ambiguity in career paths.

Formal Operational Links Between OFCOR and Extension Programmes

The structure of Training and Visits (T&V) Extension Systems creates a constant institutional demand for locally adapted technological messages to present at the regular extension meetings. If relevant information is not available, the contact farmers lose interest and the extension agents lose credibility. Several programmes in the study, including outreach programmes in the commodity programmes of Nepal, a research-extension programme in Bangladesh, and several provincial teams in Zambia, are organized in cooperation with extension to feed into systems of this type. To be effective, these kinds of links require the coordination of the larger research and extension organizations.

Table 3.1

THREE TYPES OF AGRICULTURE SUMMARIZED

Characteristics	Developing	Types of Agriculture Developed Green Revolution	Resource-Poor
Main locations	Industrialized countries, and specialized enclaves in the Third World	High capacity irrigated and rainfed areas in the Third World	Rainfed areas with steep slopes, uncertain rainfall, poor roads, and other limitations
Diversity of farming environment	Relatively low	Moderate	Very high
Major type of farmer	Highly capitalized family farms and larger units	Large and small farmers	Small farm households
Security of access to land, water, timber, etc.	High	Variable	Low
Types of farming systems	Highly specialized	Some diversity	Highly complex
Yield stability	Moderate risk	Moderate risk	Very high risk

(Continued)

Table 3.1 (Cont.)

THREE TYPES OF AGRICULTURE SUMMARIZED

		Types of Agriculture Developed	
Characteristics	Developing	Green Revolution	Resource-Poor
Use of purchased inputs	Very high	High	Low
Current production as percentage of sustainable yield	Too high	Near the limit	Moderate
Priority policy goals	Reduce surpluses, stabilize farm incomes	Sustain production for urban consumption and for export	Increase output for home consumption and for sale
Priority research goals	Reduce costs with sophisticated technology	Stabilize sustainable yields	Raise output with baskets of diverse technologies
Need for farmers' participation in R&D	Linkages already well established	Moderate need	Very high need

Adapted from Robert Chambers,"The Second Half of the World Food Conference and the Priority of the 'Third' Agriculture". Paper presented at the ILEIA/E.T.C. Workshop: "Operational Approaches in Participative Technology Development in Sustainable Agriculture," Leusden, Netherlands, April, 1988.

APPENDIX

The Case Studies

Avila, M., E.E. Whingwiri and B.G. Mombeshora. Zimbabwe: a case study of five on-farm research programs in the Department of Research and Special Services, Ministry of Agriculture.

Budianto, J., I.G. Ismail Siridodo, P. Sitpus, D.D. Tarigans and A. M. Suprat, Indonesia: a case study on the organization and management of on-farm research in the Agency for Agricultural Research and Development, Ministry of Agriculture (AARD).

Cuellar, M., Panama: un estudio de caso de la organizacion y manejo del programa de investigacion en finca de productores en el Instituto de Investigacion Agropecuario de Panama (IDIAP).

Faye, J. and J. Bingen, Senegal: organization et gestion de la recherche sur les systemes de production, Institut Senegelais de Recherches Agricoles (ISRA).

Jabbar, M.A. and M.D. Zainul Abedin, Bangladesh: a case study of the evolution and significance of on-farm and farming systems research in the Bangladesh Agricultural Research Institute (BARI).

Kayashta, B.N. and S.B. Mathema, Nepal: a case study of organization and management of on-farm research.

Kean, S.A. and L.P. Singogo, Zambia: a case study of organization and management of the Adaptive Research Planning Team (ARPT), Ministry of Agriculture and Water Development. OFCOR case study No. 1, International Service for National Agricultural Research, The Hague.

Ruano, S. and A. Fumagalli, Guatemala: un estudio de caso la organizacion y manejo de de la investigacion en finca en el Instituto de Ciencia y Tecnologia Agricolas (ICTA). OFCOR case study No. 2, International Service for National Agricultural Research, The Hague.

Soliz, R., P. Espinosa, and V.H. Cardosa, Ecuador: un estudio de caso de la organizacion y manejo del programa de investigacion en finca de productores (PIP) en el Instituto de Investigaciones Agropecuarios (INIAP).

BIBLIOGRAPHY

Ashby, J.A., Methodology for the Participation of Small Farmers in the Design of On-farm Trials, Agricultural Administration, Vol. 22, pp. 1-19, 1986.

Biggs, S.D., Determinants of Agricultural Technology: Generation and Diffusion in Developing Countries, Norwich, U. K., School of Development Studies, University of East Anglia, 1986.

Biggs, S.D., Resource-poor Farmer Participation in Research: Experiences from Nine National Agricultural Research Systems, International Service for National Agricultural Research (ISNAR), The Hague, 1988 (In press).

Biggs, S.D. and E.J. Clay, Sources of Innovation in Agricultural Technology, World Development, Vol. 9, No. 4, pp. 321-336, 1981.

Chambers, R., The Second Half of the World Food Conference and the priority of the 'Third' Agriculture, Paper presented at the ILEIA/ E.T.C. Workshop: "Operational Approaches in Participative Technology Development in Sustainable Agriculture", Leusden, Netherlands, 1988.

Ewell, P.T., Organization and Management of Field Activities: A Review of Experiences in Nine Countries, OFCOR Comparative Study Paper No. 2, International Service for National Agricultural Research (ISNAR), The Hague, 1988.

Merrill-Sands, D., Farming Systems Research: Clarification of Terms and Concepts, Experimental Agriculture, Vol. 22, pp. 87-104, 1986.

Merrill-Sands, D., ISNAR's Study on the Organization and Management of On-farm, Client-oriented Research in National Agricultural Research Systems. In: International Workshop on Agricultural Research Management, International Service for National Agricultural Research (ISNAR), The Hague, 1987.

Merrill-Sands, D. and J. McAllister, Strengthening the Integration of On-farm Client-oriented Research and Experiment Station Research in National Agricultural Research Systems: Management Lessons from Nine Country Case Studies, OFCOR comparative study paper No. 1, International Service for National Agricultural Research (ISNAR), The Hague, 1988.

CHAPTER FOUR

AN OVERVIEW OF AGRICULTURAL EXTENSION SYSTEMS

William M. Rivera 1/

This overview of agricultural extension systems has four major sections: 1) extension definitions and systems, 2) an analysis of national arrangements of agricultural extension services, 3) a critical review of recent developments relating to extension and 4) recommended priorities for the future.

EXTENSION DEFINITIONS AND SYSTEMS

Extension Definitions

Comparisons of agricultural extension systems are useful for at least four reasons:

1. the academic value of such comparisons,
2. their value in administrative decision-making,
3. their relevance to policy makers, and
4. their ultimate benefits for farmers and the rural community.

1/ Associate Professor, Department of Agricultural and Extension Education, University of Maryland, College Park, Maryland.

Comparative analysis helps sort out various factors that complicate discussion of extension systems. These include:

- -- varying concepts, definitions and terms,
- -- the interdependencies of the extension subsystem and other subsystems in the agricultural development process,
- -- the variety and multiplicity of systems,
- -- the complexity of key internal and external factors that influence the success of extension, and
- -- the lack of available programme and economic data on extension.

These factors have been discussed in detail elsewhere (Rivera et al. 1988), but the various definitions of extension require review. There are at least three definitions:

1. Agricultural performance -- extension viewed only in terms of improving production and profitability of farmers,
2. Rural community development -- extension viewed as serving to advance rural communities, including the improvement of their agricultural development tasks, and
3. Comprehensive nonformal community education -- extension viewed as a provider of nonformal agriculturally related continuing education for multiple audiences: farmers, spouses, youth, rural community, urban horticulturists, etc.

In some cases, as with the U.S. and Canadian extension systems, all three definitions operate within one extension organization. In developing countries, however, most systems tend to adhere more strictly to agricultural production services. Indeed, the current tendency among policy makers internationally is to narrowly view extension as the enhancement of the flow of knowledge between research and farmer -- in brief, as technology transfer.

Extension Functions and Purpose in Agricultural Production Institutions

Definitions of extension are often based on the operations of particular national extension systems. These definitions reflect the organizational choices made as a result of the functions the extension system is asked to perform. Thus, assumptions about how extension functions shape our definition of it.

For some, extension's purpose is, purely and simply, to deliver technology. Others broaden its functions to include education and problem solving. Still others add feedback, and involvement in adaptive research as well. Some argue that extension should also provide institutional technology, by helping farmers to organize into associations and other forms of cooperative activity.

Information delivery uses many channels of communication from the extension service to clientele. This implies the employment of agricultural information specialists (AISs) as well as agents, or village extension workers (VEWs).

Educational programme delivery involves the preparation of nonformal educational programmes, which are then delivered by extension specialists and agents to upgrade the knowledge, skills and attitudes of clientele. Again, the agent is seen as partially dependent for support on other individuals; often these are subject-matter specialists (SMSs).

Problem-solving refers to the expertise, knowledge and skills needed to solve individual and group problems arising on farms and in farm homes and families. In this instance, the agent is assumed to be well trained in farm management and, therefore, able to engage in more than technology information delivery.

Information feedback is a problematic function, and one that gets more discussion than action. This function requires, in part, that agents listen to farmers talk about their needs and their reactions to new practices and technologies and report on these to their superiors as well as to researchers. This implies that agents will engage in a diagnosis of farmer needs.

Adaptive research is currently seen as a controversial responsibility of extension. Should extension agents be assigned to engage in adaptive research projects? And should, therefore, researchers be instructed to include agents when adaptive research is undertaken? Is involvement in adaptive research one of extension's functions? Certainly, this function is not only crucial but supports the functions involving dissemination of information, knowledge and problem-solving. Indeed, it contributes to the professional enhancement of extension services.

Finally there is the question of extension's role in promoting institutional technology among farmers and rural workers. The success of farmer associations in developing agricultural production in Taiwan and the importance of the nation-wide agricultural cooperative federation in Korea are reasons for wanting to include organizational skills as part of the agenda for extension. There is agreement that the function is a significant part of the agricultural development process but no agreement as to the ways in which extension should function.

The Extension Function in Various Institutional Settings

Many assume that agricultural extension in developing countries is the responsibility of the Ministry of Agriculture. In reality, there are a variety of institutional settings that incorporate some or all extension-type functions.

Extension-type functions may be primary to an agency or organization, as with the agricultural extension service; secondary, as with private firms and cooperatives; or supportive, as with credit institutions and supply and marketing agencies. An incipient literature is developing on extension-type activities in different agricultural institutional settings, such as marketing extension (Narayanan 1986).

Types of Extension Systems

Several authors have described extension systems (Oxenham and Chambers 1978, Orivel 1981, Pickering 1987, Ray 1985, Weidemann 1987), but there are contradictions among the typologies and some confusion of terms. Pickering differentiates the following systems:

- the commodity-focussed approach in extension, designed to facilitate the production of a single crop,
- the community development-cum-extension approach, incorporating a broad definition of the functions of the extension agent, which tends to dilute the agent's specific agricultural extension responsibility,
- the technical-innovation centered approach, set up to transfer technology from outside to the farm, sometimes specifically to sell a number of technical innovations,
- the training and visit system (T&V) approach, organized to serve the farmer by mobilizing the extension system, as well as its linkage with the research system, through regular visits by agents to farmers and by regularizing of agent training, and
- the rural animation approach, associated with francophone Africa. This approach involves participatory rural development with specialists working directly with small farmers to develop, test and demonstrate improved agricultural technology.

Weidemann (1987) enumerates a similar set of models for extension delivery. These include: conventional agricultural extension, the T&V system, university- organized agricultural extension, the commodity development and production system, integrated agricultural development programmes, integrated rural development programmes and farming systems research and extension programmes.

Ray (1985) identifies three categories, which he labels as models: the directive (top-down delivery systems), the participatory (systems involving farmer participation), and the contractual model (systems where farmers contract directly with public agencies or private companies to receive extension services). Ray then recommends a fourth -- the hybrid model - which, incorporates elements of the first three models.

Lele (1975) places extension broadly under two major rubrics: the take-it-or-leave-it approach, where farmers are free to accept or reject development innovations and the contract farming approach. In the latter, farmers are granted a license to produce certain commodities on the condition that they use a particular innovation and follow project guidelines set down by the extension organization.

Oxenham and Chambers (1978) and Orivel (1981) draw attention to another approach. They categorize extension systems according to representative participation -- the touchstone of the Taiwanese Farm Information Dissemination System (FIDS), wherein both local government and farmer associations are involved in controlling the system. Axinn (1987) goes even further, categorizing approaches by point of control: the delivery approach (top-down, supervisory, and supply-driven) and the acquisition approach (bottom-up, farmer-determined, participatory, and demand-driven).

In the following analysis, based partially on the above, I arrive at a classification of four basic approaches to extension. These are then illustrated by different sets of extension systems and types of relationship to farmers. This analysis is illustrated in Table 4.1.

As this paper is concerned primarily with production extension systems, discussion of top-down delivery services, participatory acquisition systems and contract farming systems is most germane. Although valuable for information dissemination, rural development approaches usually have purposes beyond those of agricultural extension.

With the top-down delivery services, farmers may, as Lele (1975) states, take the information proffered or leave it. Indeed, research shows that it is one thing to become informed about an innovation and quite another to become convinced of its utility (Katz 1961, Rogers and Shoemaker 1971). In participatory acquisition systems, farmers have influence over extension delivery. This approach has a particular virtue and may be called the "take it or demand different packages or programmes" approach. In the contract farming systems approach, farmers must take the information, or lose their contract; i.e., they must take it or else!

Farmers' Degree of Influence on Extension Systems

The question of degree of farmer influence on extension raises questions of power and control. Should farmers' associations have control, or at least some influence, over public extension activities? Should the goal of sustainable development include client influence over extension systems?

Multiple Extension Systems

Production extension systems are frequently assumed to be one unified extension system. This is not usually the case. There are often multiple systems of agricultural extension employed by a variety of agencies and programmes within the same country. Production extension services may exist independently for crop, livestock, forestry and other agricultural products. Patterns differ from country to country, but rarely is only one agency in charge of all production extension activities. Research, extension and training for a single commodity are usually based either in a separate ministry or in an export-oriented commodity board.

A different organizational pattern prevails in some francophone West African countries. In those cases, the Ministry of Agriculture is responsible for planning and coordinating agricultural development but maintains only a few central services. The ministry gives responsibility for research and extension to parastatal organizations or special project implementation units that often operate free of central government regulations concerning personnel recruitment, contracting, budgeting, procurement and other matters.

The preceding observations point out that extension systems are often not only multiple within the public sector but include other, separate services outside the public sector. While some extension specialists (Benor et al. 1984) argue for unifying production extension systems -- at least within the public sector -- it is, nevertheless, obvious that agricultural extension in the public sector is generally multiple and involves a conglomerate of enterprises.

NATIONAL ARRANGEMENTS OF AGRICULTURAL EXTENSION SERVICES

The following broad examination of national extension arrangements is intended to be useful in understanding the challenges that face policy makers concerned with improving national extension services or changing national arrangements for a extension, or both.

Table 4.2 presents an overview of national arrangements, combining these with extension system approaches and their relationship to farmers and sets the stage for later discussion of recent developments involving extension.

Table 4.2 enlarges on Table 4.1 and provides a working outline of approaches and systems and their farmer relationships, as well as national arrangements of extension. What is lacking is knowledge of the relative effectiveness of each of these arrangements.

Public Sector Extension

At least three distinct government arrangements of the main production-oriented, public extension agencies in Africa and Asia have been identified. These are the sectoral governmental service type, the subsectoral parastatal intervention type, and the unified service with mobilization of local resources type (Blanckenberg 1984). In essence, these three arrangements operate either under, or in close connection with, ministries of agriculture.

1. Sectoral governmental service. Two sub-types are distinguished in a sectoral government service type of extension. In one, the Department of Agriculture type, the Department operates extension field services under the Ministry of Agriculture; in the other, the extension organization comes directly under the Ministry of Agriculture or another ministry. The sectoral government service type is the most common extension service arrangement in Africa and Asia and is the type usually referred to as conventional or traditional extension.
2. Subsectoral parastatal intervention. In the subsectoral parastatal intervention type of extension service, a ministry contracts with one or more parastatals. The focus is on one or a few commercial crops. This type is found in francophone West Africa with parastatal societies built on the French colonial CFDT (Compagnie Francaise de Developpement Textile) model. The societies generally have a high degree of autonomy and are responsible to the Ministry of Agriculture, which limits its concerns to overall planning, coordination and regulatory work.
3. Unified service with mobilization of local resources. The unified service type of public sector rural development and extension organization has developed in Korea and Taiwan. Its characteristics include mobilization of

> resources at the local and regional level, strictly decentralized extension programming, and development work entrusted exclusively, or almost exclusively, to one service (Blanckenberg 1984).

In Taiwan, the extension system depends on the Department of Agriculture and Forestry and operates a number of Agricultural Improvement Stations. Extension work, however, is primarily organized by farmers' associations, which wield considerable influence. Those associations carry out purchase, sale and banking functions as well as extension responsibilities, with the objectives of improving the situation of the farming population and developing the rural economy.

Private Sector Extension

In developing countries, extension functions are carried out by private enterprises that essentially are of three varieties:

1. for profit enterprises which include domestic enterprises (large farm estates, domestic firms, and cooperatives) and multinational enterprises and their subsidiaries
2. membership associations, e.g., farmers associations, and
3. non-profit organizations, e.g., non-governmental organizations (NGOs).

Domestic and multinational firms, despite certain differences, share a common market orientation. They all seek to make a profit by selling goods and services. Membership associations share an interest in profit making but are not set up for that purpose. NGOs, in general, are non-profit.

For Profit Enterprises

1. For profit large farm estates and domestic companies Large farm estates and domestic companies dominate agriculture in most developing countries. The domestic companies involve relatively small commercial enterprises and lack the level of management capacity common to large corporations, such as the multinational enterprises and their subsidiaries. The domestic operations include a wide variety of agricultural production, supply and marketing organizations.

A common feature of farm estates and domestic companies -- and other agribusiness operations -- is contract farming. Contract farming is generally limited to high-value cash crops, such as coffee, tea, cacao, sugar, tobacco and cotton, and involves technical extension or technology transfer. In the interest of maintaining a constant supply of quality products, corporations provide a package of services including improved seeds, fertilizers, pesticides, mechanical services, and a large staff of technicians for in-field supervision of farmers. Farmers are obligated to accept extension recommendations as part of the contractual relationship.

2. For profit cooperatives Agricultural cooperatives are share-issuing, private bodies. They help their shareholders acquire new knowledge and skills, which can be used to increase agricultural productivity. The quality of extension work is usually higher than in national extension services because high-value crops make it possible for cooperatives to pay better salaries and hire top-level technicians.
   In cooperatives where hierarchical social and economic relationships of the patronage system exist, egalitarian functioning is severely limited. Thus, it may be a mistake to expect that cooperatives will quickly bring about rural development. Nevertheless, cooperatives and their extension function represent an important contribution to the gradual development of group organization and the strengthening of rural communities in low-income, as well as in the more-developed, countries.
3. For profit multi-national enterprises Multi-national enterprises operate in one of two ways. They may act independently of the public-sector production system or serve as parastatals for the public sector -- as with the CFDT-type arrangements in West Africa. Many governments decentralize their planning, decision making and management functions through the parastatal arrangement. In such cases the Ministry of Agriculture delegates responsibility to an international or domestic company that undertakes, inter alia, the extension functions.
   Parastatal companies, such as the Kenya Tea Development Authority and the Sudan Gezira Cotton Project, as well as the West Africa CFDT-type companies, are often based on contract farming with compulsory extension. Non-performing farmers -- those who do not adopt cultivation specifications -- are excluded. This undoubtedly

contributes a great deal to the high rates of new technology adoption, but farmers relinquish a certain flexibility as well as the ability to respond to changing markets through diversification. Moreover, because the producers pay for the services, there is a "perform or else" climate.

Non-profit membership associations

1. Farmer associations These are membership organizations financed by fees from members. They differ from cooperatives, where members are share-holders. In some cases, as in Korea and Taiwan, farmer associations serve as a partner in the process of agricultural development. They are participants in decisions involving the implementation of adaptive research as well as field extension activities. Farmer associations, along with domestic enterprises and cooperatives represent major local resources. Policies to mobilize these resources indicate, in part, the extent to which the country and its culture can permit farmer control, as well as the extent to which the philosophy of the national arrangement is geared to farmer independence and education for self-reliance. It also raises questions that touch on a country's short-term and long-term extension priorities. Is the first priority to develop a country's production or its institutional capacity? Should public extension services be concerned about transferring production technology or institution technology? These questions will be addressed later.
2. The Non-profit role of the NGOs NGOs have a special role to play in developing countries. They have neither the financial resources nor the staff to compete with governments, international organizations or private companies. They do have, however, vital human resources, usually young people, willing to take on the difficult and sensitive task of working with rural communities at the base level. International officials continually refer to the role of NGO's in assisting marginal and subsistence farmers and in providing help with basic needs among rural populations. Important work is being carried out, especially in disadvantaged areas, by NGOs. Unfortunately, these rural development extension activities tend to lack continuity.

RECENT DEVELOPMENTS

Recent developments indicate a new environment of questioning and exploration regarding extension systems and the transfer of knowledge. These include:

-- the focus on private-sector provision of extension,
-- the privatization of certain public extension systems,
-- the trend among large farmers to bypass public extension services,
-- the effort by certain research institutions to provide what has been referred to as "frontline extension",
-- the development of new mechanisms for linking research and extension,
-- the search for participatory methods, and
-- the experimentation with hybrid research and extension systems.

These developments reflect concern with three major issues:

-- the control and purpose of agricultural extension (i.e., its ownership and orientation),
-- the right mix of extension systems for clientele conditions (i.e., public, private and mixed-type systems), and
-- the measures for improving systems so as to dynamize the extension process (i.e., how to change systems for the better, either through structural reform or functional improvements).

Many questions should be asked:

-- Do public extension systems require major changes or would managerial improvements suffice?
-- Which is the first priority, extension system development or farmer organization? Is the question one of transferring production technology or institution technology?
-- What is the main concern? Are adoption rates the prime consideration? Or is participation in and influence on the agricultural development process -- including the research and extension institutions -- the main concern?
-- Which extension systems work best? What are the best organizational structures for promoting equitable, progressive development in Third World countries? What are the best managerial modalities for service delivery?

Should systems - and their training programmes - be distinguished according to whether they promote message delivery, offer farm management services or assess and diagnose needs?

-- Which national arrangements work best? Should the Ministry of Agriculture provide extension field services? If so, should it provide the services in conjunction with farmer associations? If not, should it delegate responsibility to a parastatal? Or should it privatize the public system, thereby becoming a fee-based service, or transfer extension services to private farmers' associations while maintaining only regulatory functions? Or should there be a mix of public and private extension services, with each sector serving different clientele?

These are a few of the questions confronting policy makers and others concerned with agricultural extension and the best means of fulfilling its function. The following sections review recent developments in light of these questions.

The Focus on Private-Sector Provision

Public-sector extension, although not without some success, has generally been disappointing in transferring improved technologies from research to the farmer in developing countries. Extension institutions and programmes exist in virtually every developed and developing country and yet, in the latter, the coverage of farm families is still limited. The effectiveness of government extension services as a viable technology diffusion method has been seriously questioned by some developed countries and donor agencies.

Private-sector extension is one alternative to the conventional public agricultural extension system. The private sector is diverse, consisting of individual farm enterprises of all sizes, agricultural input industries, agro-service enterprises, processing industries, marketing firms, and multinational corporations or their subsidiaries, as well as cooperatives. The latter should not be overlooked in planning national arrangements for extensions.

In examining the private sector, a 1985 study of credit, input and marketing services (USAID 1985) concluded that public, private and mixed delivery systems each have advantages in particular situations. Public institutions are preferable when benefits are diffuse, public policies need changing or when increased economic equity is a primary goal. Mixed public-private entities work best when agricultural services require not only intensive, responsive and flexible management, but also political influence to achieve programme objectives. Strictly private firms perform best when flexible management and direct and continuing interaction with farmers are what is needed.

The USAID study also concluded that private-sector extension can serve as an important supplement to government extension systems for certain groups of producers under certain circumstances. However, private firms cannot substitute for public agencies when the policy and regulatory environment is poor, when target populations are remote, when infrastructure is lacking or when production is mainly basic food commodities grown by subsistence farmers.

The Privatization of Public Extension Systems

Some countries are moving quickly towards complete privatization of extension, as in The Netherlands and New Zealand. However, it may be too soon to promote such a move in developing countries. Public-sector services are critically important in the rural areas of many developing countries. Private-sector organizations can play a predominant extension role for providing production inputs, handling some outputs (i.e., commercial crops and commodities) and for certain farmers in particular areas.

The By-Passing of Public Extension Systems

In countries such as the United States and Canada, large and highly specialized farmers often by-pass agricultural extension services and go directly to universities or research agencies to obtain farm management information. This has caused critics and policy makers to question the need for a public extension function. The situation, however, appears to highlight the need for public extension services, because it is less feasible for middle- and small-scale farmers to contact researchers or take advantage of private sources of knowledge.

Some suggest that the by-passing of extension by some farmers is part of the natural evolution of the changing importance of extension and research at different knowledge levels. Figure 4.1, originally developed by Dr. Joao Barbosa of The World Bank, Recife, Brazil, illustrates that concept.

I argue that Figure 4.1 is operative only in certain cases. In reality, extension services are continually important even to educated farmers with expert knowledge. On the other hand, research often proves to be of immediate importance even to farmers with less knowledge.

Research Services for Extension: Frontline Extension

In some countries, the extension function is integrated into research organizations. One example is India's Lab to Land programme, an outgrowth of the Indian Council of Agricultural Research (ICAR). Prasad (1985) refers to the direct training and advice provided farmers by the ICAR/Agricultural University-assisted KVK Farm Science Centres in India as "frontline extension" -- extension information and knowledge provided directly from research specialists to farmers.

Thus, there is a role for research in training farmers -- especially those who can afford to travel to the research centres. However, this role for research appears to be a supplement rather than a substitute for other extension services.

New Designs for Linking Research and Extension

In some small countries agricultural extension is integrated with research in an effort to maximize the use of existing personnel. In other countries extension and research operate from within the same institution, with advisory committees overseeing the management of their linkage.

Evaluation of these new designs and mechanisms for linking extension and research is not complete. However, such experimentation is a welcome change from the often separate efforts of these functionally interdependent institutions.

The Search for Participatory Methods

The United Kingdom's Overseas Development Institute (ODI) has taken particular interest in the degree of farmer participation in research and extension. They refer to conventional models of agricultural research and extension as "characterised by direct transfer of technologies developed on-station to the farmer" and "unlikely to produce technologies suited to the diverse, complex and risk-prone environments in which many LDC farmers are located." According to ODI (1988), this has led to "numerous efforts to develop alternative, more participatory approaches."

The Farming Systems Research and Development (FSR/D) approach, recently altered to the FSR/E (Farming Systems Research and Extension) approach, is supported by USAID. FSR/D has encouraged great expectations that have yet to be realized, in part because of the lack of a true extension component.

The FSR/D, according to Shaner et al. (1982) is an on-farm research and development approach to farming systems which comprises the following tasks: (1) site selection (target and research area selection); (2) diagnosis (problem identification and development of the research base); (3) design (planning on-farm research); (4) research (on-farm research and analysis); and (5) extension (extension of results).

According to ODI (1988), "the problem is how to spread the costs of participatory research over a larger number of clients without detriment to the high degree of relevance achieved through participatory approaches." This statement is a contradiction in terms. The promise of FSR/D is a high degree of relevance, which, by implication, requires individualized, adaptive research. To "spread the costs of participatory research" implies that FSR/D would no longer be individualized and participatory.

The Hybrid Approach

Various authors (Denning 1983, 1985; Moris 1987, Ray 1985) consider the hybrid approach to be best suited to meet research-extension needs. Denning initially proposed the possibility of integrating farming systems research with T&V-type agricultural extension systems. Moris also discusses a possible hybrid composed of the top-down T&V system and the more bottom-up FSR/D system.

According to Moris, hybrid management systems (such as the proposed merger of T&V and FSR/D) may yet become "the breakthrough in extension productivity for which resource starved Third World extension agencies have been searching." According to Seepersad (Rivera et al. 1988), the hybrid model has already been used with positive results in the Caribbean Agricultural Extension Project (CAEP). CAEP, funded by USAID and administered jointly by the University of West Indies and Midwest Universities Consortium for International Activities, uses multidisciplinary teams to conduct Rapid Reconnaisance Surveys, a technique associated with FSR/D. Such surveys were completed for seven countries in 1986 and out of those emerged various recommendations, including one to merge FSR with T&V extension.

A hybrid, however, is formed by taking two independent, diverse, distinct varieties each of which has its own rational existence prior to hybridization. Extension and FSR should not, however, have independent existences. On the contrary, FSR is the adaptive aspect of research with which extension should be involved. Therefore, the term hybrid is misleading. While this point may seem subtle, it is important because the assumption underlying the hybrid proposal is that extension is not seen as being involved in adaptive research efforts. Herein lies one of the major problems in discussing extension, and it reminds us that involvement in adaptive research efforts is not necessarily considered an extension function. And yet, we know that technology is more likely to be appropriate, and therefore adopted, when adaptive research is done cooperatively on-farm by researchers and extensionists.

RECOMMENDED PRIORITIES FOR THE FUTURE

I recommend four directions, or priorities, for the future. They fall under the headings of development policy goals, national arrangements for institutional change, extension agency development, and social science research requirements.

Recommendations:

1. Clarify policy strategies for agricultural and research extension development

What is sorely needed at the top is vision -- clear views of national strategies for development and the interconnectedness of overall, regional and sectoral goals. Only then can there be the national will to advance with purpose. Agriculture is crucial to all countries. Does policy reflect this? How? Are research and extension seen to have mutual goals, while at the same time recognizing that they may individually embrace separate professional responsibilities?

Government must also recognize the need for the continual upgrading of the management skills of senior-level extension officials. This priority cannot be left to the extension agency because senior-level officials will not fall under its authority. Such a mandate must come from the highest level.

There is a need for agricultural extension management training at all levels (Venkataraman, 1986; Rivera 1987). This is a priority especially for top officials -- commissioners and secretaries who operate at the policy and budget levels -- as well as directors of extension. It is important for policy makers to become more cognizant of the realities of the farmer and the low and middle-level extension officers serving farmers, and of the complex management skills required to make extension's effort successful.

2. Develop the right mix of national arrangements for research/extension/development

Determinations must be made at the highest level concering the right mix of production development organizations, (public and private), the role of each, and the coordination between them. Drawing on Table 4.2, what is the best arrangement to enhance the development of research and extension? Is public system cost-recovery possible? How can these arrangements be made so as to mobilize local resources and develop them?

The development of local resources should be a major priority for all countries, but especially those considered to be least- or less-developed, because they are the ones that must learn how to gain world-market strength. Even though government may not wish to intervene directly in the development of cooperatives and farmer associations, it may do so through extension services.

Governments may be attracted by the idea of rapid agricultural development by multinational private-sector enterprises but in the long run it will be the domestic domain, including the small farmers, that must develop. The mobilization of local resources along with local government in developing extension has been shown to be socially and economically effective in countries such as Korea and Taiwan.

Extension services controlled by unified government and farmer associations -- as a result of the mobilization of local resources -- is a type of public-sector rural development and extension organization that has developed in Korea and Taiwan. The services in these two countries differ in some respects, but they share certain characteristics: mobilization of resources at the local and regional level; strictly decentralized extension programming, and development work entrusted almost exclusively to one service. In Taiwan, the government service depends on the central Department of Agriculture and Forestry and operates Agricultural Improvement Stations. The main extension work, however, is organized by the farmers' associations, which are cooperative organizations of considerable influence in rural development. These associations carry out purchases, sales and banking functions as well as extension responsibilities, with the objectives of improving the situation of the farming population and developing the rural economy.

This unified service mobilizes farmers to participate in the operation of the extension service and represents a form of joint system under the responsibility of both government and farmer associations. Lionberger and Chang (1981) cogently argue that the latter arrangement may hold out the best alternative for equitable, progressive development in third world countries.

3. Promote extension agency development

Public extension has suffered continuing attacks because in some cases its services are moribund or inefficient. Extension must, as has been the case in many developed countries, (a) re-define its strategy and goals (based on national priorities), (b) develop its staff through management and programme training -- with clarity as to what these skills should be (based on its purposes), (c) ensure that programme operations amount to at least 15-20% of the budget, (d) provide incentive systems to encourage motivation among staff members, (e) provide hardware and software for extension agents (in some cases, this will include housing as well as transportation facilities and audiovisual materials), and (f) support farmer involvement in programme development and evaluation.

To date, most management training courses and workshops have been aimed at extension agents and mid-level management. These have included senior-level extension directors and assistant directors, but the management training has focussed on service management and not on organization management relevant to extension.

Support for farmer involvement in extension and research is also an important priority. A participatory approach to extension may be the most equitable and efficient in the long run. It recognizes that planners and technocrats may not have sufficient familiarity with highly diverse agricultural needs, constraints and potential in every region and subregion in the country to enable them to design sustainable national programmes appropriate to all farmers and regions. Thus, farmers are viewed as participants in decision making, rather than passive recipients.

The degree of farmer influence on extension is a major question. Should farmers' associations, as in the cases of Korea and Taiwan, control extension activities? Certainly farmers should have some say in the systems that presume to serve them. While it may be too much to ask that extension take on the burden of promoting democracy, it is a tool - or set of functions - that may enhance democratic mechanisms.

4. Support social science research requirements

There is a critical need for data collection, comparative study and evaluation in agricultural extension. Ideally, programme effectiveness studies as well as economic cost-benefit analyses should be undertaken. Extension programmes may be highly effective in helping to resolve existing production and income-generation targets, even if economic cost-benefit analyses, with their focus on production yields, do not show impressive gains.

Lack of adequate data has presented a problem for extension policy makers and providers for years. Systematic, continuing data collection is needed through monitoring and evaluation systems. While monitoring and evaluation are often spoken of in the same phrase, they require differentiation, even though monitoring information may contribute to evaluative studies. Monitoring is essentially a management supervisory tool. Evaluation, as the word implies, is a judgemental research method aimed at helping with decision-making.

CONCLUSION

Three major underlying concerns are revealed in this overview. The trend towards more efficent extension systems, especially towards private-sector extension, indicates the concern with economic viability. Also donors, as well as individual countries, are questioning the value of backing single models of extension. Consequently, there is a growing concern for situation specificity. Finally, it appears that linkage is being placed high on the list of major concerns for extension success. The importance of system interdependence is being reflected in several ways: through technology system analytical frameworks (Swanson 1986), in major interdisciplinary studies (such as by the International Service for National Agricultural Research) and in new national arrangements for research-extension linkage.

The new environment of questioning and exploration regarding extension systems and the transfer of knowledge appears to be a healthy development, but answers are not yet clear and will probably be found in small rather than large segments. Although concerns with economic viability have highlighted the efficiency and effectiveness of private enterprise and put public extension defenders on the defensive, it is obvious that the private sector is only interested in certain low-risk, high-yield situations and, therefore, is not sufficient to develop the agricultural sector in its entirety in developing countries.

In addition, the trend of the last two decades to develop models of extension appears to be coming to a close. The principles of extension, not systems, are coming into the limelight. Indeed, the concept of situation specificity argues for diversified systems. While incipient, it appears that national arrangements fostering new policies and institutional designs in this regard are finally coming into being. The critical need now is for analyses that show the relative effectiveness of these and other national arrangements for extension.

Table 4.1

EXTENSION APPROACHES AND SYSTEMS AND THEIR RELATIONSHIP TO FARMERS

System Approaches	Type of System	Relationship to Farmers
I		
Top-down Delivery Services	Conventional T&V University-organized Technical innovation Integrated agricultural development programmes	Take-it-or-leave-it
II		
Participatory Aquisition Systems	Farm information dissemination (Taiwan) Farming Systems Research and Development (FSR/D)	Take it or demand different packages or programmes
III		
Contract Farming Systems	Commodity development and production Commodity focussed	Take it... or else
IV		
Rural Development Extension approaches	Community development-extension cum-extension Rural animation Integrated rural developent programmes	Take-it-or turn away

Table 4.2

NATIONAL ARRANGEMENTS AND SYSTEM APPROACHES FOR EXTENSION AND THEIR RELATIONSHIPS WITH FARMERS

National Arrangement	System Approach	Relationship with Farmers
1. PUBLIC		
Ministry of Agriculture (field services)	Delivery service (non-compulsory) Arrangement #1	Take-it-or-leave-it
2. DUAL		
Ministry of Agriculture and farmers' associations (dual control over extension services)	Participatory (shared responsibility) Arrangement #2, 4b, 4c	Take it or demand new package or programme
3. COORDINATED		
Ministry of Agriculture (parastatal). May be public or private		
4. PRIVATE		
(a) for profit: domestic enterprises cooperatives multinational enterprises	Contract farming (compulsory) Arrangement #3, 4a	Take it ... or else
(b) membership: farm associations		
(c) non profit: NGOs		

Figure 4.1

THE RELATIVE IMPORTANCE OF EXTENSION AND RESEARCH AT DIFFERENT CLIENT KNOWLEDGE LEVELS

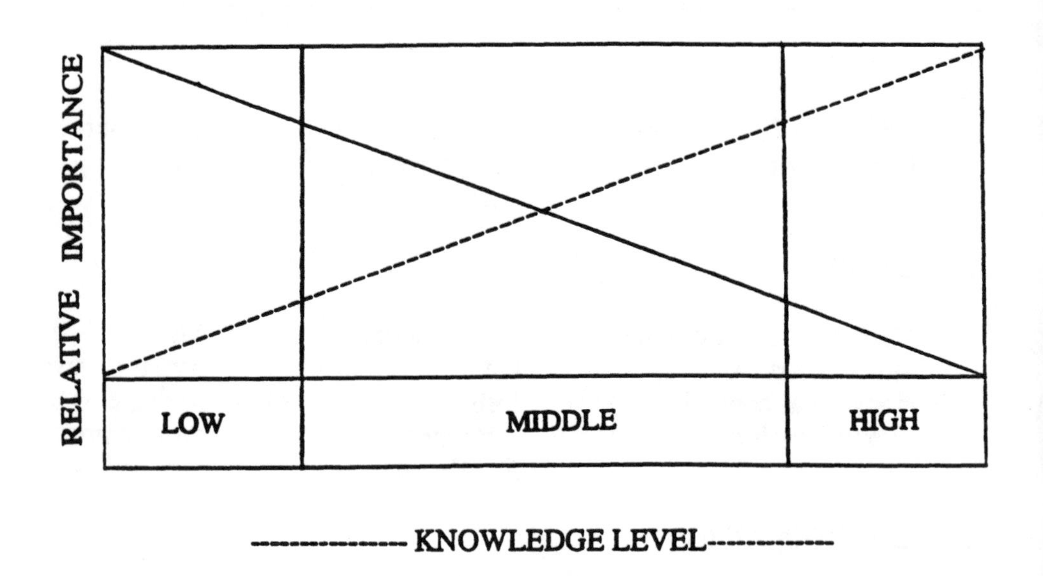

--------- Research

———— Extension

BIBLIOGRAPHY

Axinn, G.H., "The different systems of agricultural extension education, with special attention to Asia and Africa". W.M. Rivera and S.G. Schramm eds., Agricultural Extension Worldwide, Croom Helm, London, 1987.

Benor, D., J.Q. Harrison and M. Baxter, Agricultural Extension: The Training and Visit System, The World Bank, Washington, D.C, 1984.

Blanckenberg, P. von., Agricultural Extension Systems in Some African and Asian Countries: An Analysis of Country Reports, FAO, Economic and Social Development Paper, Rome, 1984.

Burbank, R. and P. Flynn, Agribusiness in the Americas, Monthly Review Press, New York, 1980.

Crowder, L. Van, "Public and Private Sector Extension in Agricultural Development". W.M. Rivera and S.M. Walker eds., Lifelong Learning Research Conference Proceedings, University of Maryland, College Park, 1987.

Denning, G., "Integrating Farming Systems Research with Agricultural Extension Programmes", Paper prepared for The World Bank Asian Regional Conference, International Rice Research Institute, Los Banos, Philippines, 1983.

Denning, G., "Integrating Agricultural Extension Programmes with Farming Systems Research". Cornea, M.M., Coulter, J.K. and Russell, J.F.A., Research-Extension Farmer: A Two Way Continum for Agricultural Development, Washington, D.C., The World Bank, 1985.

Katz, E., The Social Itinerary of Technological Change: Two Studies on the Diffusion of Innovation, In Human Organization, XX (Summer), 1961.

Lele, U., The Design of Rural Development: Lessons from Africa. Johns Hopkins University Press, Baltimore, 1975.

Lionberger, H.F. and H.C. Chang, "Development and Delivery of Scientific Farm Information: The Taiwan System as an Organiza- tional Alternative to Land Grant Universities -- U.S. Style" Extension Education and Rural Development, Volume 1, International Experience in Communication and Innovation, John Wiley, New York, 1981.

Moris, J., "Incentives for Effective Extension as Farmer/Agency Interface". W.M. Rivera and S.G. Schram eds., Agricultural Extension Worldwide, Croom Helm, London, 1987.

Narayanan, A., "A Survey of Agricultural Marketing Extension in Six Countries," Unpublished paper, Center for International Extension Development, College Park, Maryland, 1986.

Orivel, F., "The Impact of Agricultural Extension Services: A Review of Literature," Discussion paper, The World Bank, Washington, D.C, 1981.

Overseas Development Institute. Personal Communication on Farming Systems Research/Extension Meeting, London, 1988.

Oxenham, J. and R. Chambers, Organizing Education and Training for Rural Development: Problems and Challenges, The World Bank, Washington, D.C, 1978.

Pickering, D.C., "An Overview of Agricultural Extension and its Linkages with Agricultural Research: The World Bank Experiences," W.M. Rivera and S.G. Schram eds., Agricultural Extension Worldwide, Croom Helm, London, 1987.

Prasad, C., Linkages Between Agricultural Research Education and Extension in India, Indian Council of Agricultural Research, New Delhi, 1985.

Ray, H.E., "Incorporating Communication Strategies into Agricultural Development Programmes". Part I. Guidelines, II. Agricultural Communication in the Context of Agricultural Development Programmes, Vol. III. Extension and Communication Models, Academy for Educational Development, Washington, D.C, 1985.

Rivera, W.M., "India's Agricultural Extension Development and the Move Toward Top-level Management Training", W.M. Rivera and S.G. Schram eds., Agricultural Extension Worldwide, Croom Helm, London, 1987.

Rivera, W.M., J. Seepersad and D. Pletsch., "Comparative Agricultural Extension Systems," D.J. Blackburn eds., Foundations and Emerging Practices in Extension, University of Guelph, Guelph, Ontario, 1988.

Rogers, E.M. and F.F. Shoemaker, Communication of Innovations: A Cross-cultural Approach, Free Press of Glencoe, New York, 1971.

Shaner, W.W., P.F. Phillip and W.R. Schmehl, Farming Systems Research and Development: Guidelines for Developing Countries, Westview Press, Boulder, 1982.

Swanson, B.E., "Analyzing Agricultural Technology Systems: A Research Report", INTERPAKS, University of Illinois, Urbana-Champagne, 1986.

United States Agency for International Development (USAID), "Agricultural Credit, Input and Marketing Services", Evaluation Special Study Report No. 15. Washington, D.C, 1985.

Venkataraman, A., Extension Managment Training Needs, The World Bank, New Delhi, 1986.

Weidemann, C.J., "Designing Agricultural Extension for Women Farmers in Developing Countries," Agricultural Extension Worldwide, W.M. Rivera and S.G. Schram eds., Croom Helm, London, 1987.

CHAPTER FIVE

SOME ASPECTS OF AGRICULTURAL EXTENSION EXPERIENCES IN INDIA

J. L. Bajaj 1/

The World Bank's 1986 World Development Report noted that many developing countries discriminate against agriculture: "The general economic policies the developing countries have pursued have, however, limited the growth of agricultural production and hampered efforts to reduce rural poverty."

According to this line of thinking, the development strategies followed by those countries have discriminated against agriculture in an effort to promote industrialization by erecting high trade barriers. "Such strategies are intended to accelerate the shift of resources out of agriculture by lowering its profitability compared with that of industry so that agriculture is worse off than it would be in relation to domestic prices let alone to relative world prices."

The role of prices in augmenting agricultural production has been extensively discussed and emphasized in recent years. However, in the absence of non-price measures, prices alone are not likely to achieve the objective of increased agricultural production. Non-price measures, including timely access to credit, fertilizer, pesticides, water supply, transport and land systems, are equally important in contributing to

---

1/ Joint Secretary, Ministry of Agriculture, New Delhi, India.

agricultural production. Agricultural extension is an important non-price measure supporting farmers in increasing their productivity.

This Chapter describes the extension system as it has evolved in India and reviews the field operations in the context of extension activities covered under a project financed by IFAD.

INDIA'S EXTENSION SYSTEM

In the early 1950s, the need for specific agencies to implement rural development programmes in India was recognized, and a new administrative unit, the Development Block, was established as a primary unit in developmental administration. Headed by a block development officer (BDO), the Block was created to assist the development departments dealing with agriculture, horticulture, cooperatives, animal husbandry, village industries, etc. Its staff included a number of assistant development officers (ADOs) and village level workers (VLWs). The latter were assigned specific sub-units, normally consisting of a few villages. The VLWs promoted development programmes and made arrangements for inputs such as fertilizer, pesticides and credit. Although they would function as multipurpose workers, it was stipulated that VLWs would spend 75% of their time on functions related to the promotion of improved agricultural practices. Over time, the activities of different development departments increased, as did the number of such departments operating at the village level. That increased the workload of the VLW. In addition, the VLW came to be increasingly deployed in activities not connected with agriculture -- elections, census, family planning, etc. -- which reduced the time devoted to agricultural programmes. This, along with the realization that the process of modernization of agriculture required an efficient communication and delivery system, led to the introduction of the Training and Visit (T&V) system of agricultural extension in a number of states in India.

The Training and Visit System

The T&V system is characterized by the following features:

-- A unified extension system, under which all extension activities relating to agriculture are the responsibility of agricultural extension workers, over whom administrative control is exercised by the Agriculture Department.

-- Extension workers are expected to devote all their time to extension activities. They are not given responsibility for any regulatory functions, input supply, provision of credit or any activities relating to other rural development programmes. This is because of the experience that when other functions are added, extension relating to agricultural activities, per se, suffers.

-- Every extension worker is responsible for a given number of farmer households which are generally divided into groups. The village extension worker is expected to meet each group on a fixed day each fortnight for extension work. This ensures regular flow of information and facilitates supervision of extension workers.
-- Regular training is imparted to the entire hierarchy of the extension staff.
-- On the basis of the training received, extension workers transmit messages to the farmers in their area through contact farmers. In the T&V system, the contact between extension worker and the contact farmers is close and regular. The selection of contact farmers and the manner in which they convey the messages to other farmers is very important in determining the success of the programme.
-- There is a close linkage between research and extension activities.

Growth of Extension

India's extension system is now the largest in the world. About 90,000 persons are employed in extension work in 17 states that have T&V projects being implemented with World Bank assistance. In addition to those with direct responsibility for agricultural extension, a staff is employed by the Agriculture Department for reaching farmers with inputs, credit distribution, soil testing, etc.

In Tamil Nadu, the number of VLWs more than doubled between 1982 and 1987. Increases on a similar scale occurred in Madhya Pradesh and Gujarat. There is a wide variation in the proportion of expenditure on extension activities between states. This varies from as little as 0.7% in Orissa to as high as 14.7% in Bihar. The total expenditure on extension in 11 states for which detailed information is available was Rs. 797 million (US$75.3 million) in 1987-88. This constituted about 8% of the total outlay on agricultural activities in the states. This expenditure, however, relates only to extension projects financed by The World Bank. Expenditure on extension activities is often hidden in the outlays of other schemes. The World Bank estimates that Rs. 3250 million (about US$250 million) are spent annually in India on different extension projects.

A recent World Bank review of the extension programme in India found that the institutional aspects of a reformed extension system are well in place. Trained staff have been posted in districts and villages. Operational units have been established with research institutions and the extension staff receives regular training. Extension workers visit farmers regularly and there is sufficient evidence that farmers are absorbing new technology. This has resulted in a significant change in cropping patterns and increased agricultural production in several areas. The Bank report concludes that the T&V approach is a powerful force for disseminating information in rural areas. The report recognizes that there is considerable room for improving the performance of extension services. The report's authors feel, however, that problems often arise because casual observers are unable to perceive the benefits of extension and extension workers are unable to demonstrate the effects of extension through conclusive data.

The Extension Project in Haryana is one in which a positive contribution to agricultural productivity has been recognized. In a study of two districts -- one in Haryana and the other in Uttar Pradesh -- with comparable agro-economic conditions and resource endowments, it was found that the needs of the farmers had been met far better in Haryana than in Uttar Pradesh. In Karnal district, Haryana, 86% of the contact farmers and 45% of non-contact farmers were visited by VLWs in 1981-82, but in Muzaffarnagar district, Uttar Pradesh, extension workers visited only 7% of the farmers. This showed that the visits by extension workers in Haryana were more regular and as a result their contact with farmers more productive. A large proportion of contact and noncontact farmers perceived a change in the style of extension operation and held positive views about it. Contact farmers were found to have passed on the knowledge derived from extension workers to noncontact farmers. Moreover, seventy percent of contact and 36% of noncontact farmers in Karnal found that the reformed extension delivered more useful and timely information than the old system. In Karnal extension workers were recognized as the primary source of information, and other sources, (radio, publications, etc.), were of far less significance.

Extension in Uttar Pradesh Tubewells Project

Extension is a relatively recent input in Uttar Pradesh and was introduced in the IFAD/World Bank financed Public Tubewells Project. Any successful system of extension has to be integrated in the system of development administration prevailing in the area. Much of its success will depend on the support it can draw from the institutions and individuals who shape the development process. In the specific instance of Uttar Pradesh, a revealing insight into the attitude of key functionaries of the Agriculture Department and the public is provided in the statements recorded by the Commission on District Level Administration. The Director of Agriculture told the Commission that at the district level there was no mechanism for coordination between the state agricultural officers, district plant protection officer, district local officer and district soil conservation officer. They worked under their respective regional- and state-level officers. He also stated that the T&V system was outdated as it did not envisage any initiative on the part of the farmer. Most members of the public interviewed by the Commission were of the view that the number of government functionaries in the field was excessive. They were not in favour of the idea of separate Agriculture Department village-level workers for catering to the requirements of agricultural programmes. These views, though not universally held, represent some of the reservation on the reformed extension system expressed by development administrators and influential public men.

Uttar Pradesh has a population of about 110 million and a geographical area of 298,000 square kilometres. Its economy is predominantly agrarian with agriculture accounting for 56% of the state's income and employing 78% of the labour force. The annual per capita income, at Rs. 2567 (US$121) in 1983, was among the lowest in the country. The growth in income has averaged 2.2% per annum and growth in foodgrain production about 2% per annum. The increase in food grain production, mostly wheat and rice, was due to area expansion rather than rising yields; growth in oil seeds and sugarcane productivity has been slow.

Uttar Pradesh has abundant water resources. However, only about 8.5 million hectares of the total cultivated area of 17 million hectares was irrigated in 1979-80. The use of groundwater has increased steadily. The public tubewells programme is more than 50 years old. The tubewells are installed in areas not serviced by surface irrigation schemes and where the water table is too deep for operation of private wells, or in backward areas where private investment is unlikely to be made. The performance of state tubewells has not been satisfactory mainly because of insufficient power supply, incomplete water conveyance systems, inadequate water allocation procedures and unsatisfactory operations and maintenance.

Solutions to the above problems were sought through the implementation of a pilot project with International Development Association (IDA) credit of US$18 million. The project involved construction of 559 public tubewells and was designed to demonstrate and evaluate the relative merits of a number of technical and operational improvements to the design of the existing tubewell system. These improvements were aimed at reducing the water losses, protecting equipment from electrical breakdown, installing an automatic operation system, providing more acceptable water allocation procedures and ensuring better maintenance and faster construction. During the implementation of this pilot project, it was recognized that the power supply arrangements were inadequate to deliver full project benefits. The agricultural extension programme in the state was also not structured to provide necessary services to help farmers use the improved water availability.

Those shortcomings were addressed in the second phase of the project. This involved construction of 2200 new tubewells, rehabilitation of about 100 existing tubewells, implementation of agricultural support services to farmers in command areas, formation of a research and development programme to test further potential advances in tubewell technology, training of irrigation and agricultural staff and project monitoring and evaluation. The tubewells were to be located in a cluster of about 25 units and connected to a power system by separate feeder lines servicing only public tubewells, with a view to ensuring power supply of at least 16 hours a day during the peak demand period. The project covers 43 districts of the state. The agricultural development component covers all the new tubewells of phase II, as well as those constructed under the pilot programme.

Because Uttar Pradesh had no formal arrangement to deal with the special post-construction requirements of the tubewells, a separate Agricultural Development Wing (ADW) was created in the Agricultural Directorate to work exclusively for the introduction of effectively irrigated agriculture in the tubewell command area. The ADW in the field comprises three agricultural supervisors at tubewell cluster level reporting to an agricultural officer (AO) assisted by an additional agricultural officer (AAO). The charge of a supervisor thus varies from eight to nine tubewell commands. The AAO is responsible for two tubewell clusters and the AO for six clusters -- about 150 tubewells. Three deputy directors at the regional level supervise three AOs each. Administrative control at the state level is with a joint director who reports directly to the director of agriculture. Two subject matter specialists are attached to each deputy director at the regional level.

The main functions of ADWs are to provide close contact with the farmers, introduce new technologies, assist in the preparation of crop plans, assist in the organization of farmers' groups, identify the requirements and local sources for the timely supply of agricultural inputs and credit, provide guidelines to farmers in water management techniques and organize and administer farm demonstration plots. Once the majority of tubewells in a cluster have reached the full development levels assumed under the project, the cluster is handed over to the regular extension service responsible for that area and the ADW staff moves on to a new cluster.

The AOs are required to give technical advice and support, ascertain that farmers are being visited regularly, confirm that they are receiving appropriate technical recommendations, determine whether farmers are adopting their recommendations and whether their income is increasing, check that contact farmers are selected according to the guidelines, review agricultural supervisors' daily diaries and participate in farmers' group meetings.

Agricultural supervisors are required to visit every tubewell command once in a fortnight on fixed days. They select about ten farmers as contact farmers and are required to meet contact farmers as well as other farmers during their visits. Before the sowing season, a special training camp is organized in every tubewell command to familiarize the farmers with the latest developments. Agricultural supervisors organize demonstrations. During their farm visits, they are required to advise the farmers on various aspects of agricultural development, collect information on crop conditions and prepare and suggest remedial action. In addition, they listen to farmers and encourage them to discuss difficulties, conduct group meetings, note the extent to which their recommendations are being adopted and the reasons for the failure to adopt them more widely. In addition, they teach and demonstrate the recommendations learned during training in the previous fortnight. Supervisors are also required to maintain a daily dairy. They are the conduit for providing feedback to research. They also help the farmers in arranging inputs and credit and also prepare cropping and input plans for the farmers.

The project provides for training of both junior and senior staff of the ADW. The training of senior staff members covers soil-water-plant relationships, estimation of crop irrigation requirements, intensive cropping patterns in irrigated agriculture and agronomic aspects of water management, irrigation scheduling of different crops, surface irrigation methods for efficient application of water, land levelling design aspects and cost-benefit considerations. Agricultural supervisors are trained in water requirements of crops, irrigation efficiencies and irrigation scheduling, as well as planning, design and layout of surface irrigation systems, soil surveys and agronomic aspects of main crops. The staff members are provided access to agricultural universities and crop research centres. In addition, monthly training at the state headquarters and fortnightly training at regional headquarters are arranged. The purpose of the training is to update the knowledge of the extension staff and to provide constant feedback for research and other activities.

Phase II has brought irrigation and an extension service to the project area and the impact is clearly discernible. Cropping intensity increased by about 25% in the first two years of project implementation. A similar trend occurred in the Phase I project area.

Although full benefits are yet to be realized from the project, productivity has shown a marked increase. It increased by more than 25% for rice and 30% for wheat in the first two years. The cost of production has gone up, mainly attributable to increased use of inputs. The increase in the cost of production has, however, been more than offset by increased returns, which have enabled the farmer to sustain higher investments. Use of fertilizers has increased considerably, and pesticides and improved seeds have shown similar tendencies.

Irrigation and extension in this project were introduced almost simultaneously. It is, therefore, difficult to assess the impact of a single factor on agricultural production in quantitative terms. However, for a general assessment of the strength, relevance and impact of the extension service in the project area, discussions were held with supervisors in four different tubewell commands in Lucknow and Hardoi districts. Conventional tubewells in both districts were also visited for purposes of comparison. The progress and benefits of the project were also discussed with the staff of ADW at state headquarters.

Supervisors in Lucknow district have been in position in their respective areas for three to four years. The extension service in the tubewells project was initially envisaged only for three years in the post-construction stage. However, because extension is still to acquire roots and services provided by conventional community development programmes were considered inadequate, the extension services are being continued. Although Lucknow district has been included in the T&V programme as part of the National Agricultural Extension Projects, its impact is yet to be felt.

Extension has played an important role in popularizing the use of fertilizers in the project area. The farmer today is more aware of the benefits of fertilizers although he may not apply them in quantities advised. The use of pesticides has also increased although it is noticed that non-availability of working capital is an important inhibiting factor in furthering their use. In general, the farmers are vigilant against crop diseases and receive useful advice from the extension workers on curative and preventive action. Introduction of new high yielding varieties in the project area is attributable to the extension work done. Maximum impact has been evident in wheat production, followed by rice production. No significant change has taken place on farms producing pulses.

Agricultural supervisors visualize their role as giving advice to farmers on the use of inputs and techniques for increasing agricultural production. However, in many cases, the farmers feel that the visits of supervisors are of a routine nature and lack specificity and direction. The farmers are generally aware of the visit days and of the role of supervisors but have yet to develop confidence in their advice. The supervisor's main effort appears to be directed at meeting the contact farmers and ensuring that demonstration targets are met. The system of regular meetings on each visit has been done away with because of apparent lack of interest by farmers. As a result, the visits of the extension worker do not last a full day.

Examination of the daily diaries kept by agricultural supervisors found that the diaries contain little useful information. There is no record of the discussions held with farmers, or of their special problems, or of the conditions of the standing crops or any information on the outbreak of diseases. Most of the diaries contain the typical statement: "discussed farming techniques with the farmers".

Though crop and input plans are prepared for contact farmers, there is no followup to ensure that the recommendations are followed. In most cases, even when farmers have a shortage of working capital, they are reluctant to use credit for the purpose of procuring fertilizers for fear of crop failure.

The recommendations of extension workers on cropping patterns broadly follow the priorities determined at the state level. In an almost routine manner, wheat and rice are recommended in the respective seasons although some other crops can be profitably grown in this area. For example, in parts of Hardoi district soil in large areas is ideal for oilseeds; yet, no efforts have been made to encourage farmers to grow mustard. Crop selection is almost entirely a matter of a farmer's individual judgment, which is greatly influenced by personal requirements. The cropping pattern has not shown any significant change other than that attributable to the introduction of irrigation in the area.

In water management, the extension services have had very limited impact. The farmer uses as much water as he can get. Therefore, special efforts need to be made to make the benefits of controlled irrigation visible to him.

Uttar Pradesh has three agricultural universities -- G.B. Pant University at Pant Nagar, C.S. Azad University at Kanpur and N. Dev University at Faizabad -- which have been assigned the responsibility for basic and applied research within their operational areas. Adaptive research remains the responsibility of the Department of Agriculture, which presently has ten Regional Testing and Demonstration Stations. Besides, there is an Irrigation Research Institute at Roorkee and a Horticulture Research Station at Saharanpur. Central Research Institutes for sugarcane, grassland and soil conservation have been set up in the state. As these facilities were considered adequate for project purposes, no additional provision was made for investment in research activities.

In the project areas visited in Lucknow and Hardoi districts, the only link to research is through demonstration farms, which are handled casually and serve only a limited purpose. For example, application of fertilizers on these farms is based on standard prescriptions and does not take the soil characteristics into account. Soil tests are not made regularly. Even when they are, the results of these tests are available only after a considerable time gap. In fact, in some cases, the results were made available after two or three seasons. Consequently, such results are seldom useful to farmers.

Demonstrations in the project area follow a top-down approach and their usefulness is limited. Selection of crops for this demonstration is largely governed by the monetary limits for demonstration fixed by the state government. Thus potato, which is the preferred crop of many farmers because of significantly higher returns when irrigation is assured, has not been selected for demonstration because the input costs for a single crop exceed the total sanction available for all three crops. Moreover, potato is considered a horticultural activity and the extension workers who are employees of the Irrigation Department do not see any role in it for themselves.

A majority of the farmers in the project area prefer to grow mixed crops. They believe that this ensures against the vagaries of nature because the chances of survival of at least part of a crop are greater than when a single crop is planted. Yet there is little evidence to suggest that this felt need has been considered and that correct techniques and advice for raising such mixed crops have been sufficiently developed.

The fortnightly training of extension workers is not planned properly. The training period is used more in the nature of a departmental meeting at which the performance and targets of utilization of fertilizers, use of high yielding seed varieties, cropping patterns, etc. are reviewed rather than as an opportunity to train supervisors. Only the training held at the start of the sowing season is related to the crops grown in that season. The extension workers also have little interaction with subject matter specialists. The meetings with AOs and AAOs are considered sufficient for upgrading knowledge. Supervision of the work of field staff is inadequate. The AOs do not follow any fixed schedule of visits and their interaction with farmers is limited both in frequency and duration. This is partly explained due to lack of mobility. A large number of vehicles were targeted for the project but only one-third of the motorcycles provided have been purchased due to lack of financial sanctions.

Needs for Effective Extension Systems

I am of the view that the success of the extension system depends, to a large extent, on the quality of advice tendered, the availability of complementary inputs (such as irrigation, seeds, credit, fertilizer, etc.) and the quality of the delivery system. The advice given by the extension staff, in turn, depends on research linkages. These linkages are often weak. Wherever such a link is effective, extension is able to identify the problems of farmers and provide feedback for research. Successful solutions, when formulated, are then transmitted to farmers through the extension agency.

A successful research-extension linkage should have the following elements:

- Extension messages should be based on research done in the agro-climatic zones.
- It is not enough to transmit the recommendations of the research station within an agro-climatic zone. The recommendations have to be further adapted by considering the local soil condition and the availability of water, fertilizer and other inputs.
- Research trials should be undertaken in the fields for which the resource endowments of farmers, conditions of tenure and adoption by farmers of recommended practices would be relevant factors.
- An effective extension system is needed to identify the problems and communicate them to research stations.

Where extension and research have worked together, they have been effective in finding solutions to many farmer problems. In a case in Madhya Pradesh, it was noticed that despite the use of recommended practices, the yields declined. Investigations by the research station revealed that the decline was mainly due to zinc deficiency. Its subsequent application restored the productivity of the land. Many such examples have been recorded. However, in spite of notable successes, it is recognized that the research and extension linkage in India is quite weak. Some of the reasons for this have been identified as:

- Reluctance of scientists in agricultural universities and departments of agriculture to accept change in concepts and procedures.
- Lack of appreciation of each other's role by extension officials and scientists.

-- While physical infrastructure has been made available, there is little decentralization of the administration of research and extension because of insufficient institutional freedom, competency and motivation.

The major problem today is to increase the agricultural productivity of small and marginal farmers, especially in rainfed areas. Until scientists are able to develop a worthwhile technology for this, large investments in manpower for agricultural programmes will continue to be questioned. Along with technology, small and marginal farmers also lack adequate resources for using fertilizer, pesticides and other inputs in appropriate measures.

As many observers of the Indian agricultural scene have noted, the smaller farmers use land and water more efficiently than larger farmers. Investment in infrastructure has suffered because of a lack of resources available to the states. About 500 irrigation projects are estimated to be incomplete for want of funds. In a situation where water and credit are constraints and no worthwhile new technical information is available for rainfed agriculture, investment in extension can only be of marginal benefit.

It may be desirable to link extension not merely to research but also to input supply functions. The extension workers need to ensure that inputs are available to farmers, that the farmers are able to adopt such inputs and that any investment undertaken by the farmers is likely to be worthwhile.

The quality of the delivery system has been a subject of much comment. Many unhelpful critics have characterised the T&V system as "touch and vanish." If the expected results have not been achieved, it is also because the motivation of the extension staff is affected by low salaries, poor service conditions and stagnation in promotions. The T&V system addresses itself to these problems by allotting a fixed number of farmers to each extension worker and by fixing the day on which the extension worker will visit an area. Written reports have been reduced to the minimum to ensure that the VLW spends maximum time in the field with the farmers. In extension projects funded by The World Bank, provision is often made for housing and improved mobility. There is also the need to improve the communication skill of staff members. The present emphasis on one-way communication of advice from SMSs to VLWs needs to be shifted to a free flow of information in both directions. Extension workers should be encouraged to analyze the problems of farmers and convey them to technical personnel for advice. They should also be advised to take into account the constraints within which the farmers work. There should be a greater focus on discussing and analyzing the reasons for nonadoption of specific extension advice by farmers. This would enable a better insight into the problems of farmers.

The role of extension is far wider than merely communicating technical messages received from the top. Its task is to comprehend and appreciate the relationship between production, financial resources available and technological practices. The limitations and difficulties faced by farmers have to be presented to the administrative system and alternative solutions discussed with farmers. All this involves far more than conveying messages from laboratory to land.

Nevertheless, it may be stated that given its limitations, India's reformed extension system is an improvement over the disorganized and under-funded system that preceded it.

CHAPTER SIX

WORLD BANK INVOLVEMENT IN AGRICULTURAL EXTENSION

John A. Hayward 1/

For more than 20 years the World Bank has supported agricultural extension as a central thrust of its policy to promote agricultural advancement - the engine of growth for most developing countries. Since the mid 1960s the Bank (including the International Development Association) has provided about US$45 billion to support more than 1,200 agricultural and rural development projects, of which almost 500 included some form of extension input. Direct Bank funding for extension has reached almost US$2 billion for 79 countries worldwide.

The magnitude and spread of its investment funding are just one aspect of the Bank's influence on extension. Interacting with governments over long periods, direct involvement in design and implementation of agricultural projects and programmes at the field and policy-making levels, and worldwide experience with extension has made the Bank a recognized force in extension concepts. The Bank's prominence in assisting to promote extension has, however, led both to some misconceptions about its role and to resistance on the part of some borrowers to change in the extension process.

1/ Rainfed Crops Adviser, The World Bank. The views expressed are those of the author and should not be construed as official policy of The World Bank.

After a brief review of the Bank's involvement in extension, I discuss how the principles of industrial management should govern the extension process and how the process should be determined along the lines of flexibility, site specificity and economic viability of the systems employed. An assessment of how the lack of knowledge of extension's clients and lack of appropriate technology can lead to poorly designed extension systems has been made. This section is followed by an overview of the main issues that have been shown by the Bank's experience to constrain extension effectiveness. It also raises issues in the areas of policy, demand, supply and national organizations.

WORLD BANK INVOLVEMENT IN EXTENSION

A detailed review of World Bank lending for extension from 1965 to 1986 will soon be published. Briefly, that review will show that Bank-supported projects involving extension have increased rapidly over time, from one per year in the mid 1960s to 10 per year in the mid 1970s and to 34 per year since 1983. Bank lending for extension has similarly risen from about US$1 million per year in the 1960s to almost US$200 million per year in the 1980s. Investment has been greatest in Latin America and South Asia, which together absorbed 52% of extension funds. Africa attracted the greatest number of individual extension projects, mostly small and highly focussed. Brazil has been the largest World Bank borrower for extension (US$328 million) followed by India (US$279 million), Nigeria (US$138 million), Mexico (US$133 million) and Indonesia (US$94 million).

The Variety of Bank Involvement in Extension

The World Bank's support for extension has taken various forms. Some involved minor extension inputs at project level but in others the Bank worked with governments over long periods to create complete national extension systems. Two varieties of extension systems are identifiable as discrete approaches in past Bank-supported extension: the commodity specific system employed in about 12% of projects, and the training and visit (T&V) system, introduced in about 30% of the projects.

Current Bank lending for extension by geographical region is shown in Table 6.1. Almost 40% of the projects are in Sub-Saharan Africa with extension costs averaging about US$6 million for each project. Asia and Latin America now take up over 60% of Bank lending to extension, although only comprising 45% of the projects.

The T&V system, included in 65 Bank-supported projects, is particularly prominent in Asia and -- more recently -- Africa. Table 6.1 illustrates, however, that current Bank-support for extension is in no way confined to T&V alone. In Africa, Bank support manifests several different forms: in Zaire, it is provided to farmers through direct involvement with non-governmental organizations (NGO) in Zambia extension components support semi-commercial commodity companies and in Cameroon full commercial services, including extension, are being provided by parastatal agencies and cooperatives. Moreover, in Nigeria, while the main thrust of extension in area development projects is based on T&V principles, radio and video systems support in forestry projects is provided through local community and religious leaders.

In Asia, the Bank is involved in strengthening extension programmes in India, Bangladesh, Sri Lanka and Indonesia through the introduction of T&V. However, discrete extension support is being given through media communication development in China; by providing village and township extension and agricultural input service centers in Burma; and in employing progressive farmers and local retired businessmen as part-time agricultural advisors in Nepal.

In Latin America, the Bank is supporting privatization of parts of the extension service in Mexico and is closely involved with the development of extension computer systems in several countries. In Brazil, the Bank supports programmes that involve developing producer associations and cooperatives as well as encouraging more advanced farmers to employ specialist agricultural advisors.

These on-going activities illustrate the scope of extension activities with which the Bank is involved and the flexibility of its approach to different countries' extension needs.

Table 6.2 presents the 1988 Bank lending for extension by the agricultural sub-sector and shows that Bank interest in extension stretches across the full range of agricultural activities. Area development and irrigation extension projects together make up more than half of extension project interventions, and 62% of T&V projects are included in those subsectors. The principal reason for this concentration of interventions is worth noting. Whereas extension in specific commodities, such as livestock and fisheries, usually adopts a campaign approach operating in parallel to general extension services, the introduction of extension to regional areas and to irrigation schemes generally involves the long-term development of an effective rural service for which a properly designed T&V system is particularly well suited.

BANK LENDING FOR TECHNOLOGY GENERATION AND DISSEMINATION

The Bank currently supports 27 projects directly concerned with developing national research or national extension systems or a combination of the two. The historical divergence of research and extension services in many countries has compartmentalized them into separate responsibility orbits. Unfortunately, this has led to differential government funding and to a situation in some countries where research has been promoted far ahead of extension and, in others, where extension has become the dominant service. These distortions lead to agencies growing apart. Eventually each agency attempts to compensate for perceived weaknesses in other parts of the knowledge system and embarks on programmes for which it is neither responsible nor technically equipped; thus, research takes on extension and extension does research.

The Bank's recognition of these problems and its support of attempts to overcome them are exemplified in many countries where coordinated programmes of extension and research are being developed. This is particularly apparent in India, where the National Agricultural Research Project has encouraged state agricultural universities to identify discrete agro-ecological zones and to plan research according to zonal technical needs. Meanwhile, a Bank-supported National Agricultural Extension Project in each state concentrates on organizing T&V extension systems within zones. Zonal committees, consisting of farmers, extension staff and research scientists, meet twice a year to agree on seasonal work plans. A similar arrangement is now being fostered in Zimbabwe under the National Research and Extension Project. There the Committee for On-Farm Research and Extension, chaired in alternate years by research and extension, must review and approve the funding of all proposals for on-farm trials and demonstrations. In Bangladesh, the Bank has attempted to foster coordination between agricultural research, extension and training by means of joint review missions and joint meetings with the services involved. Similar exercises in Pakistan, Kenya, Nigeria, the Philippines and Mexico serve to illustrate the Bank's commitment to the concept of strengthening the knowledge system rather than injecting funds into isolated parts of the system.

Obviously, estimating the number of farmers who will be influenced by a World Bank-funded extension activity is extremely difficult and questionable. Making estimates of the external benefits to the system is virtually impossible. However, an estimate of the number of farm families directly involved in World Bank-supported extension projects has been made for 200 of the 218 projects (Table 6.2). These projects interact with some 32 million families or about 160 million individuals. The people who benefit directly or indirectly from such support will be many times this number.

THE WORLD BANK APPROACH TO EXTENSION

The distinction between the extension process and extension systems designed to carry out that process is far more important than mere semantics and has a direct bearing on the Bank's approach to extension and to its interactions with governments over extension issues.

The essential feature of a process is that it is a dynamic, changing procedure aimed at a particular goal or objective. A system involves an organized entity, which at any time is carrying out the functions of the process. The goal of the extension process is principally the improvement in the well being of rural communities. The process of extension can be undertaken using many interlocking, complementary systems -- for example, rural development extension systems and private sector commodity improvement systems. The essential point is that there is no one correct way of organizing extension. Extension systems, to achieve the goals of the extension process, must be dynamic and flexible and must be constantly questioned as to their appropriateness and effectiveness. This is a basic tenet of the Bank's approach to extension.

The Bank recognizes the dynamic nature of extension, as shown by the wide variety of its interventions in extension support. Unfortunately, however, this flexibility of approach has been overlooked because of misconceptions over the principles and details of the T&V system.

The World Bank and T&V

The Bank is associated, through the pioneering work of Daniel Benor, with the design and promotion of the T&V system, which is now used in more than 40 countries. Although the system was widely adopted, its progress was frequently hindered by misunderstanding or ignorance of the fundamental principles on which it is based and confusion between those principles and implementation details.

T&V is a management system that applies basic organizational principles to operations aimed at using a relatively unsophisticated and often poorly motivated workforce to change the behaviour of geographically scattered, diverse individuals. The first descriptions of the system (Benor and Harrison, 1977; Benor et al. 1984) illustrated its underlying principles by means of specifically detailed examples, which unfortunately have become identified with the system itself. Thus, in the minds of many, T&V involves specified staffing patterns, fixed extension worker-farmer ratios, visits to contact farmers at fixed intervals irrespective of the message or season, and even the concept of contact farmers teaching other farmers. These are misconceptions of the system, which have become engrained in T&V lore.

To dispel the impression that the Bank is rigid in its approach to extension, I can do no better than quote Daniel Benor (1987), the designer of T&V:

"There can be no one system of extension suited to all farming communities. The variation in agro-ecological conditions, socio-economic environments and administrative structures is such that one system cannot be expected to suit all conditions. To be successful, the T&V system must be adapted to fit local conditions. However, the flexibility that enables successful adaptations to be made in the system does not allow for adaptation of its basic principles."

The basic principles that Benor sets out are:

1. T&V extension systems require: a professional service with a full-time trained staff, supported by resources required to perform their professional functions;
2. a single line of command where staff are technically and administratively responsible to one authority;
3. staff effort to be concentrated on extension activities with staff members performing clearly defined and monitorable tasks;
4. time-bound work and training programmes including regular farm-visit schedules;
5. field and farmer orientation with special reference to meeting farmers on their own fields;
6. regular and continuous training at all levels to up-grade professional skills, and
7. a procedure for ensuring a two-way flow of information between research, extension and farmers.

These 7 points are the principles of the T&V system and are drawn from the fundamentals of effective management. The details of implementation of these principles will differ between countries, between regions and even within one region over time. In other words the details are site specific, whereas the principles are universal, and should not be compromised.

Need for Modifying the Details of T&V Extension

The T&V management system is one way of organizing resources to carry out the process of extension. It has proved to be effective in circumstances where extension was virtually non-existent or where the existing system had lost sight of fundamental principles.

The impact of restructuring extension along T&V lines, which has brought about improvements in agriculture, has resulted in the dogma that any modification to the T&V system would be detrimental. In other words, T&V can be constrained by its own success. The problem stems from extension designers and implementers concentrating on extension as an end in itself and not a means to an end. As Lawrence and Lorsch (1967) stated, "instead of seeking relationships between organizational states and processes and external environmental demands...most organizational research and theory has implicitly, if not explicitly, focussed on the one best way to organize in all situations."

The problem, of course, is that the essential organizational requirements for effective performance of one task under one set of conditions may not be the same as those for other tasks with different circumstances. It is critical, therefore, to examine any extension system in relation to its goals and to question whether modification is needed. The T&V system is no exception to this rule and Bank-supported extension projects are continually modifying the details surrounding the system to accommodate local circumstances. The recent move in several countries to promote extension workers' interaction with farmers' groups rather than with individual contact farms is a case in point.

The need for careful evaluation of the effectiveness of any change is obvious. The problem is that instruments for measuring the impact of change on the knowledge system, as well as on farm productivity, are not readily available. Evidence of increases in agricultural production are notoriously difficult to measure. Even the most careful and expensive crop-cut estimates of production are open to serious question. Measuring adoption rates is a more straightforward exercise, but the results are usually confounded by many factors and sorting out the impact of extension alone is questionable. The actual impact of change on the extension system itself is even less quantifiable and requires sensitive management awareness plus a willingness to reverse decisions on innovations that prove ineffective.

EXTENSION SYSTEMS AND MANAGEMENT THEORY

Examining extension systems for their effectiveness points to the need to analyze what Lawrence and Lorsch (1967) call the differentiating dimensions of an organization. The dimensions are: goal orientation, time orientation, interpersonal orientation and structural orientation. These somewhat daunting descriptions conceal important criteria for any review of extension.

Goal Orientation

Goal orientation looks at the particular objectives of extension systems, extension managers and extension workers. Clarity of goals is fundamental to designing effective systems, yet policy makers, budget directors and extension planners rarely start by establishing goals. It is obvious that designing, funding and implementing an extension system to increase the well being of rural communities is very different from designing a service to control cotton pests, yet there is a tendency to treat all goals with the same extension prescription. The campaign approach to bringing about change in agriculture requires a very different system than the approach that provides a sustainable rural service.

Time Orientation

Time orientation recognizes that different segments of a system have different time frames. Clearly, farmers have a time frame that spans a season or two. An extension worker's time orientation may be fortnightly or monthly. Adaptive research has a medium time-frame outlook, whereas the perspective of basic research will be long-term. Accommodating the different time frames in an extension system is a major management task.

Interpersonal Orientation

Interpersonal orientation relates to the style of operating procedure adopted by the extension organization. This may be authoritative and bureaucratic -- run on military lines -- or consultative, where expertise is brought in from outside the system in the form of research workers or farmers, or participatory, where decision making is based on the continuous collective interaction of all concerned. The style of interpersonal relationships in an organization should be largely determined by knowledge of clients and knowledge of technologies.

Structure Orientation

Structure orientation refers to the reporting relationships between parts of the organization, to the control mechanisms employed and to aspects of accountability and performance monitoring. In a campaign, the system may be highly structured with junior staff reporting to senior staff at fixed schedules and in a fixed format. In informal structure, actions cannot be pre-determined and the reporting format is necessarily loose.

These four dimensions, which should be considered in designing or modifying extension systems, are directly related to the degree of certainty surrounding the knowledge of clients and the knowledge of technology appropriate for those clients. This relationship is shown in Table 6.3. The matrix indicates that where agricultural technology is known and where the type of farmer to use the technology is also known, then a more authoritative style of extension with precise targets and time frame can be employed. Campaigns and most commodity extension systems fall into this category.

At the other extreme, where neither appropriate technology nor farmer characteristics -- their demands, preferences, objectives, etc. -- are known, a fully participative approach to introducing change is essential. This involves searching for basic and applied technology, working with farmers to determine their needs and long-time horizons and, in particular, establishing a more informal type of operation. It is obvious that a campaign approach would be quite inappropriate where farmer characteristics and technology are not known.

In the intermediate situation, where either a potential innovative technology or client characteristics - but not both - are known, an intermediate type of extension system is required. This would necessitate a more consultative goal orientation: searching either for client information or for appropriate technology, and testing the technology with clients to achieve a matching situation, medium-term time horizons, and a differentiated structure where some parts may be formal and others informal. The need for a differential system structure lends itself to a combination of a delivery and feedback system working in combination with a type of farming- systems-research approach.

Table 6.3 tends to oversimplify a complex dynamic situation reflecting many interactions of clients and technology. Some technology is known to be appropriate for particular clients, whereas, in other cases, client needs cannot be satisfied. The situation changes as farmers' socio-economic situations change and as technology becomes available. The critical point is the need for constant fine-tuning of the extension system to reflect changing circumstances.

Technology packages frequently are designed on the assumption that these reflect client socio-economic needs. Therefore, in those situations, a bureaucratic extension campaign, in accordance with management theory, should be mounted to achieve precise goals. One important implication of the matrix in Table 6.3 is that if, through a false sense of knowledge of either clients or technology, the wrong system of extension is developed, then it has little chance of achieving its goals.

A further important implication of this view of extension design is that if a system, such as T&V, focusses only on the delivery side of its function, then it operates as a campaign approach. This implies that both technology and client needs are known. If the technology is inappropriate or farmers' needs differ from those perceived, then the delivery side of the T&V- structured system will fail. This emphasizes the essential feature of the feedback function of T&V, which is of equal importance to the delivery function. Feedback is not simply an appendage to the main delivery system but a critical part of the system itself -- a response to the uncertainties inherent in agriculture.

Feedback is often described as the weak spot of the T&V system. This view, however, reflects more on the quality and training of the extension staff and on poor linkages between extension and research than on the system itself. Feedback of information on the performance of innovations and on farmers' reactions -- flowing from the field to research -- is built into the T&V system in regular meetings with research staff. But, recognizing and articulating farmers' problems is an acquired skill, which requires careful training of the extension staff. To date, most training programmes have focussed almost exclusively on the delivery function and have thus distorted one of the fundamental principles incorporated in the T&V system -- the two-way flow of information between research, extension and farmers.

T&V AND FARMING SYSTEMS RESEARCH

There has been a tendency to characterize T&V and farming systems research as two different systems of extension, and there has been much debate over the advantages and disadvantages of these approaches. It is important to realize, however, that T&V and farming systems research should be integral parts of the same system.

The matching of technology to farmers' needs and circumstances is, of course, a serious concern among research institutes, development planners and extension workers worldwide. It is possible to cite many examples of research workers producing technology geared to a perceived clientele that is subsequently found to be inappropriate for use by the majority of farmers. This issue has been particularly detrimental to the needs of smaller traditional farmers who have tended to be at the bottom of the research priority list. The problem has stemmed from a belief that agricultural advancement lies in the hands of advanced farmers and that, in order to develop, all smaller farmers should adopt advanced technology. There is now growing awareness that this belief is untenable.

The challenge is to train extensionists to determine and rank farmers' problems and to train researchers to develop technology to ease those problems. Undertaking diagnostic surveys is therefore a cornerstone of the agricultural knowledge system and both research and extension have complementary parts to play in the exercise. The Bank supports this essential cooperation between agricultural research and extension and diagnostic survey work has been incorporated into extension and research programmes in several countries, including Pakistan and India as well as countries in Africa and Latin America.

That T&V research and farming systems research can be complementary is brought into focus by reference to Table 6.3. Where there is a mixture of certainty and uncertainty or where either client characteristics or technology - but not both - is known, then a differential structure of extension is appropriate. This is precisely the mixture of goal orientation, time orientation, interpersonal orientation and structure orientation exemplified by juxtaposing farming systems investigations with a T&V system of delivery, training and feedback. In these circumstances of uncertain knowledge, either T&V research or farming systems research in isolation would be inefficient and counterproductive. Together they constitute a powerful extension system.

The difficulties of managing a differentiated system of extension should not be underestimated. Managing a campaign, or even the delivery side of T&V, is relatively straightforward. Where, however, management is weak or inexperienced it adopts authoritative, bureaucratic procedures to compensate for lack of initiative, lack of commitment and lack of understanding among staff. It is argued, of course, that in many countries staff capability at all levels is weak and that incentives to perform well are almost non-existent. These are serious issues; ones that greatly reduce the effectiveness of extension. But falling back on a military style extension system to overcome weaknesses in staff is almost certain to produce an unsustainable extension service. Achieving goals with less than optimal resources is the mark of good management. The implications for careful training of staff and extension managers are obvious.

ISSUES IN EXTENSION

The issues of management capability, staff capacity and incentives are but part of the serious problems confronting extension worldwide. The World Bank's experience in working with government extension services has revealed, however, that there are rarely unique issues associated with specific country situations. There are, of course, specific details that pertain to a particular country, region or client characteristics, but these details invariably fall within a framework of areas of generic issues.

The Bank's experience suggests that these areas can be grouped within a matrix of issues comprising the government policy environment, the demand side of extension, (which is particularly associated with client characteristics), the various supply systems, and the organization of extension at national levels. These issue areas are related to three important criteria: strengths and weaknesses of linkages, situation specificity, and sustainability.

This matrix of issue areas is illustrated in Table 6.4, which itemizes some of the most important issues and puts them under criteria headings. The issue list does not pretend to be exhaustive and extension implementers may well add their own issues or transfer specific problems to other parts of the matrix. The compartments of the matrix are intimately linked. For example, it is not possible to consider the provision of budget resources -- a national organization function -- separately from government awareness of the value of extension, or to separate development goals and national strategies from the question of whether an extension service should be centralized or decentralized. Nevertheless, the matrix performs an important function -- it breaks down the multifarious problems of developing effective extension, which at times appear overwhelming, into conceptually manageable segments. It also provides a guide to where problems might occur and focuses the attention of the many agencies, both public and private, formal and informal, on the area of responsibility where intervention may be needed.

It is possible to elaborate on the Bank's experience in any issue area shown in Table 6.4. However, in the context of the over-riding theme of this paper, I have selected one or two issues to illustrate how the matrix can assist in focussing attention on the traditional poor farmers.

At the policy level a major issue relates to national development goals and strategies. If the goal of one country is to increase agricultural exports, another's is growth with equity, another's is poverty alleviation, and yet another's is rapid economic growth, then the research, extension, education and training systems may find themselves pulled in quite different directions. One of the major reasons for the gap between research and extension in many countries is that the goals of the two services are often totally different. Research may be aimed at increasing the efficiency of agricultural production, whereas extension may be aimed at spreading technology so as to equalize opportunities. The clients of these systems and approaches are very different and the problem arises not because of weaknesses in the systems per se, but because of weaknesses in coming to grips with the socio-political problems of joint extension and research goal-setting.

Within the demand issues area (Table 6.4) it is obvious that the education levels of farmers -- in the broadest sense -- have major implications for the style of extension employed. Education levels determine whether farmers can take on their true role in the research-extension-farmer continuum; whether they can interact effectively or form supportive associations; and whether they can begin to influence the knowledge system. If the poorest farmers are not taking part in extension, it is easy to blame extension staff for not operating equitably and effectively. But this overlooks one point. Often, the poorest smallholders have no education and consequently do not have the confidence to interact on a face-to-face basis with government workers. Experience shows that it is only when farmers gain a basic standing in society, usually brought about by formal or informal education or by specialized knowledge, that they can interact effectively as partners in the extension process. The relative social standing of farmers and extension field workers can be a major barrier to the articulation of the poorest farmers needs.

In this discussion of the demand side issues, it is important to distinguish between the clients and the beneficiaries of extension. In the narrowest sense the clients of the outward delivery system are farmers. However, it is often forgotten that the clients of the inward (feedback) delivery system are research workers, input suppliers, credit managers and data analyzers. Research scientists are therefore part of the supply and demand side of the knowledge system.

Furthermore, the beneficiaries of extension are the community at large and in particular farm household members. Again the implications of this beneficiary recognition must be borne in mind. If the impact of agricultural extension on households is that children move out of farming, then the effect on future household opportunities and on the needs and style of extension itself should be taken into account in the design of future extension programmes.

As regards the supply side issues, these were touched upon above in the discussion of the differences between campaigns and participatory systems of extension, in comments on the T&V system and the importance of systems management. The interaction of knowledge levels and their influence on extension style, whether controlled or unstructured, is especially important in dealing with the rural poor. It is frequently in that critically important social area that knowledge of both beneficiaries and appropriate technology are at a minimum, and, therefore, where the need for long-term, participatory extension is essential.

As a final example from the matrix, the impact of national organization and the degree to which knowledge systems should be privitized present serious issues for the poorest members of the farming community. The roles of the public and private sectors in agricultural extension are factors in development. Current trends, particularly in parts of Europe, Asia and Latin America, favour moving towards privatization of agriculture, and there are obvious advantages in this perhaps inevitable trend. However, two critical concerns surround both the public and the private debate: Will private extension services deal effectively with poor farmers who appear to present little commercial promise? And, will reliance on predominantly private extension lead developing economies more rapidly towards environmental degradation? The answers to these questions must lie in a balanced mix of public and private involvement in extension, in which entrepreneurs gradually take up the commercial opportunities, while public extension retains equity and regulatory goals.

The dynamics of this mix of public and private extension will, of course, differ between countries and over time. Management of the transition from public extension to a mixture of public and private is the responsibility of policy makers who need to set long-term goals for rural development and to be aware of the implications -- for members of society and particularly for the rural poor -- of achieving these goals.

Other issues that bear on effective support for extension for small farmers can be determined from the matrix, but discussion of each could occupy a paper in itself. Under sector linkages it is obvious that access to input supply (including credit) and access to markets is directly linked to extension's interaction with small farmers. Also, effective linkages between national systems and policy makers is important in creating awareness of the contribution of small farmers to the national economy. The costs of failing to provide extension services to small farmers must be at the forefront of policy decisions.

As concerns situational specificity, individual country or regional characteristics should be taken into account, particularly regarding farmer characteristics, social systems, staff capacity and the degree to which extension is either centrally managed or has the opportunity to respond to specific local problems.

A major issue under the sustainability of systems is the profitability and cultural acceptability of technology. Issues of quantifying social benefits and of the advisability of subsidizing inputs or prices are central to this discussion and have a direct bearing on the effectiveness of extension.

SUMMARY

I have presented the World Bank's commitment to the place of extension in the development process and have indicated some of the ways in which the Bank supports extension worldwide. The Bank views extension as an interacting constituent of development; as development progresses so must extension progress. Flexibility and adaptability are key words in building extension systems, and I have emphasized that there is no such thing as a correct system. Holding rigidly to any one extension system is as unwise as dogmatically adhering to a specific crop variety.

I have outlined how four dimensions of management theory have an important, often overlooked, bearing on extension system development; how establishing goals and ensuring that extension workers at all levels work towards those goals is fundamental to success; how time horizons of participants in extension must be recognized and managed; and how operating procedures and reporting structures should not be ends in themselves but should be geared to the state of knowledge about the existing technology and of extension's many clients.

In this regard, I have emphasized that T&V and farming systems research should form complementary parts of the same system. In drawing on Bank experience I have categorized issue areas under policy, demand, supply and organization, and have cross-referenced these to sectoral linkages, to the specifics of particular situations and to the sustainability of extension. Use of the matrix (Table 6.4) can focus attention on specific issues and separate them from the general problems that constrain extension.

Table 6.1

WORLD BANK ON-GOING REGIONAL LENDING FOR EXTENSION ACTIVITIES, MAY 1988

Region	Countries	Projects			Extension costs (million US $)	Bank/IDA lending for extension (million US $)
		Involving extension	With T&V system	With other specified involvement		
Sub-Saharan Africa	32	82	33	28	492.1	184.5
Asia	14	56	16	20	384.3	220.4
Europe, Mid-East and North Africa	11	27	9	17	393.3	123.8
Latin America and Carribean	12	27	7	24	642.8	264.4
Total	69	221	65	89	1,912.5	793.1

Table 6.2

WORLD BANK EXTENSION LENDING BY SUBSECTOR, MAY 1988

Subsector	Projects			Extension costs (million US $)	Bank/IDA lending for extension (million US $)
	Involving extension	With T&V system	With other specified involvement		
Agriculture (general)	7	2	3	139.7	25.8
Agricultural credit	4	0	1	28.8	9.8
Area development	101	27	38	759.4	322.5
Fisheries	2	0	0	6.1	3.0
Irrigation & drainage	34	12	11	74.7	29.8
Livestock	11	0	6	42.5	18.2
Agro-industry	9	2	4	14.8	6.6
Perennial crops	9	4	4	29.8	11.3
Research & extension	27	16	12	770.3	347.9
Forestry	11	2	8	14.4	7.3
Other than above	3	0	2	32.1	10.5
Total	218	65	89	1,912.5	793.1

Table 6.3

THE INTERACTION BETWEEN TECHNOLOGY AND CLIENT KNOWLEDGE AND ITS IMPLICATIONS FOR EXTENSION STYLE

Agricultural technology / Extension client characteristics Socio Economic Setting	KNOWN (High degree of certainty)	UNKNOWN (High degree of uncertainty)
Known (understandable and relatively simple)	1. Authoritative, bureaucratic approach	1. Consultative approach
	2. Performance-target oriented	2. Testing orientation
	3. Precise time requirements	3. Medium-term time horizons
	4. Formal structure	4. Differential structure
Unknown (complex)	1. Consultative approach	1. Participatory approach
	2. Testing orientation	2. Search orientation
	3. Medium-term time horizons	3. Long-term time horizons
	4. Differential structure	4. Informal structure

Table 6.4

ISSUE AREAS IN AGRICULTURAL EXTENSION

Issues related to: / Issues involved with:	Sector linkages	Situation specificity	Sustainability
1. Policy	With: - development goals - national strategies	Regarding: - importance of agriculture to national economy - government commitment	- Economic, social, and political benefits and costs
2. Demand	Farmer linkage with: - research - input supply - markets - education	Regarding: - farmer character-istics- - education levels - participation - social systems	- Financial pro-fitability - cultural needs
3. Supply	Extension link-ages with: - research - input supply - markets - farmers	Regarding: - system manage-ment - leadership - capacity - monitoring - controlled - unstructured	- cost effec-tiveness - evaluation
4. National organi-zation	Linkage manage-ment between: - policy makers - commerce - other sectors - knowledge systems	Regarding type of organization: - centralized - parastatal - public-private - commodity focussed	- budget resources - management capacity - regional cohesion

BIBLIOGRAPHY

Benor, D., "Training and Visit Extension: Back to Basics". Agricultural Extension Worldwide, W. Rivera and S. Schram Eds., Croom Helm, London, 1987.

Benor, D. and J. Q. Harrison., Agricultural Extension: The Training and Visit System, The World Bank, Washington, D.C., 1977.

Benor, D., J. Q. Harrison and M. Baxter, Agricultural Extension: The Training and Visit System, The World Bank, Washington, D.C., 1984.

Lawrence, Paul, R. and Jay W. Lorsch, Organization and Environment, Harvard University Press, Boston, 1967.

CHAPTER SEVEN

INVESTMENTS IN AGRICULTURAL EXTENSION

Michael Baxter 1/

How the considerable recent investments in agricultural extension in developing countries can be most effectively utilized and their impact sustained is an ever-increasing concern. One reason for this has been the difficulty in isolating the impact of these investments. Given the nature of agricultural extension and its existence as an integral part of a broad agricultural knowledge system, this difficulty is understandable. At the same time, however, the concern has been augmented by the fact that both public agricultural extension services and the governments of which they are a part are often unclear about the role and function of extension, and of the policy and infrastructural support needed for effective extension. The issue of sustainability of extension investment, particularly for poor small farmers, should be reviewed from the perspective of the general extension policy framework and extension's infrastructure and technological base, which is what this Chapter aims to do briefly.

1/ Chief, Agricultural Unit, The World Bank, New Delhi. The views expressed are those of the author and do not necessarily coincide with those of The World Bank.

POLICY FRAMEWORK FOR AGRICULTURAL EXTENSION

In developing effective agricultural extension systems, a basic consideration is the national policy framework within which they operate. National agricultural policies normally indicate production targets and priorities, but rarely do they elaborate on the organizational structures and objectives necessary to implement this policy nor, especially, the incentives needed to encourage farmers to act in a manner supportive of national goals. Detailed guidelines are not the concern of national policy, of course, but unless the general parameters are established and promoted at the highest political levels, the likelihood of their effective enumeration and implementation at other levels is remote (Baxter et al. 1988).

National agricultural development policy should indicate the intended contribution of extension to agricultural development and to a country's particular agricultural development strategy; moreover, it should show how they are interrelated institutionally and in the manner of their implementation. It is not necessary for national policy statements to detail the mode of extension, but basic parameters should be established. These should include its professional and technical orientation, basic work responsibilities and management principles, criteria by which its effectiveness should be monitored, ways in which extension contributes to national policy formulation and implementation, and the basic institutional arrangements linking it with other developmental services such as agricultural research, input supply, agricultural marketing and processing, and social programmes.

The need for such a policy framework, particularly for one that endures, is seldom questioned among people involved in extension and field-level development activities. Why is this the case? Aside from a possible belief that such detail is not the stuff of national policy, there are probably two main reasons. One is ignorance: policy makers, even in agricultural ministries, may forget about extension in the hurly-burly world of more urgent national priorities. A second more probable reason, is that policy makers are confused over the nature and role of extension. Being often far removed, from the field realities of farmers' options and decisions, they may attribute extension's analytic and dissemination responsibilities to research or other agencies. They may not be convinced that extension's primary function concerns agriculture, rather than comprehensive social development, or that nowhere has modern, high-output farm production been sustained without a technically trained and focussed agricultural extension service. The difficulty of effectively organizing and managing an agricultural extension service, the multiplicity of methods and potentially involved agencies and the problems of attribution to extension of a specific impact on production are other considerations that encourage many a wise person to steer away from linking extension to national policy.

Extension staff members unfortunately often contribute to this situation. Policy making is not confined to national levels, particularly when it is seen as establishing developmental objectives and the principles by which these will be attained. At each level of operation, extension managers and staff should be aware of extension's explicit objectives and their roles in attaining these. They must be able to link their activities directly to their ultimate objective -- improving the incomes and livelihood of all farmers through direct advice on production and farm management and by influencing other farmer support services (in particular, research), to improve their provision of relevant, timely support to farmers. The potential contributions of different extension agencies -- the public (government) service, supply and marketing cooperatives, private input and marketing organizations, agricultural universities, private companies and consultants, voluntary organizations, radio and television programmes -- should be established, and the ways they can complement each other agreed upon. Appropriate individual performance targets and extension service at each level need to be agreed upon and then monitored. Of course, many of these necessary steps depend on national policy, but such guidance is likely to develop only when there is some clarity of objectives and activities within the extension service itself. This need for an internal sense of direction and mission becomes more necessary and influential over time because the national policy framework and the extension organization should be continually modified to reflect changing national priorities and agricultural and institutional achievements and experience.

While government can be blamed for giving inadequate attention to extension, extension itself often has much to account for in this regard. The assignment of ineffective leaders to the extension service -- non-specialists, administrative officers, poor performers, people in need of promotion, non-field or non-farmer people, and so on -- may be identified as a prime cause of weak extension, as indeed it is likely to be because no institution can remain effective for long with weak leadership. But extension staff members themselves often allow, even encourage, this situation to develop by not adequately identifying and promoting their role and achievements in the context of national priorities. Fortunately, it is a situation that may be readily rectified provided extension focusses on the basics -- its clientele, the farmer and other agricultural support services.

In most countries, farmers -- even poor small farmers -- are a significant political consideration. Governments are generally concerned that farmers produce in order to supply non-farm populations with food and raw materials which can also be exported. Moreover, they are concerned that rural unrest does not weaken government authority. If extension serves its clients well -- helps them achieve greater incomes and to adapt effectively to the ever-changing market economy in which all significant farming populations are involved to some degree -- they will be extension's greatest supporters. Just as farmers demand well-trained, productive local extension agents, so too can their support be used to demand effective management and leadership at all levels. It is not unusual, unfortunately, for extension staff to forget that their work makes little sense unless it serves farmers. In such circumstances, they inevitably become seen, both by farmers and by government, as increasingly irrelevant to agricultural production, and indeed rather as a service suitable for odd-jobs, such as the distribution of technically and even financially irrelevant subsidies and the performance of non-agricultural monitoring and publicity functions. Similarly, a failure to give research and other farmer-support services (even though they are often its chief competitors for funds) comprehensive and analytic feedback on farmers'conditions and needs is equally debilitating. If those services don't receive such feedback, or pressure to perform, they not only see no need to support extension, but they will often attempt to undertake extension functions. The occasional cry from concerned extension staff for better government understanding and backing, unsupported by either strong farmer or support service confirmation, inevitably has little credence whatever its innate merits.

From the foregoing, three ways stand out in which the enduring government commitment necessary for effective agricultural extension can be established. One is for extension staff at different levels to be clear about their objectives and role, to be monitored according to them and to be held accountable for meeting them. A second is for extension to keep in view its role -- to serve the farmer and farmer support services. Realistically informed of extension's activities and aims, farmers are fair evaluators of extension's performance; Moreover, satisfied farmers make their support for extension known. Similarly, support services that feel the need for extension in the light of extension's contribution to their work will bolster extension.

Finally, even while extension's contribution to agricultural production may be difficult to quantify, we all know examples of good extension work. It may be the promotion of effective soil and moisture conservation measures or of less water-demanding crops in time of water scarcity, or of giving successful direction to research to focus on an ignored cropping system or more economically realistic production recommendations. Such examples should be identified and analyzed, and policy makers put in contact with the farmers involved. In many governments, senior officers in influential ministries (such as Finance and

Planning and even Agriculture) have little direct experience with the beneficiaries of the programmes they can so drastically influence.

EXTENSION INFRASTRUCTURE

Even though each extension service must be designed to take account of local cultural, agro-economic and administrative conditions, and consequently will have particular infrastructural needs, there are adequate general guidelines available from experience to help establish effective services. The fundamental guideline is, of course, that extension's infrastructure should be designed so that it can achieve the service's objectives. Obvious as this is, it often does not happen -- as in the case with the field agent who is tied to his extension-supplied office rather than being in the field, or the training system that is divorced from farm reality. In this section, guidelines relevant to major elements of the extension infrastructure are summarized.

Organization and Management

The basic principle of successful organization and management of extension at all levels is that it should be conceived as the simplest possible and then further streamlined. Whenever feasible, various technology-based, farmer-support programmes need to be combined into one service: staff and budgets should be merged and duplication in reporting and establishing activities removed. Any function or regulation that does not contribute directly to strengthening extension's field operations should be reviewed. If it is inappropriate to extension or not a priority activity, it should be either transferred from the service or stopped. Reporting channels need to be short and direct. No report should be produced that is not read or reports a situation that is better observed directly. Supervision of any activity should take place where and when the activity is being performed. Training cannot be supervised by reviewing accounts of topics covered or participants present, just as extension field activities cannot be supervised from field agents' workbooks (diaries) or agent-reported adoption rates.

Extension often suffers from muddles. The work expected of a functionary at any level must be clearly delineated, be reasonable in amount, relevant to the agricultural extension profession and backed by appropriate support (e.g., transport, technical backstopping, training and equipment) and be implemented on a known and monitorable schedule. Within these guidelines, however, adequate room must be allowed for staff to experiment and adapt to local conditions. Provided staff members understand their objectives and the parameters of their responsibilities and mode of work, they can be innovative without disrupting their basic duties.

Staff

A basic guideline is that there is no immutable number of staff members for an extension service. There are seemingly countless variables in determining ideal staff numbers: the field service structure; strength and relevance of alternative extension systems (the media, input and marketing companies and cooperatives, welfare services, agricultural universities, voluntary organizations, etc.); rural settlement patterns, cultural considerations, and the interests and sophistication of farmers; extension staff quality; strength of agricultural research and relevant technology availability; communication systems; extension objectives; uniformity- complexity of farming systems and so on. Moreover, each of these variables changes constantly.

Within this apparent confusion, however, there are some relatively unchanging considerations. First, the fewer the staff members necessary to fulfil extension's objectives, the better it is from the managerial, financial and quality points of view. Second, some model agent-farmer ratios are required for planning purposes, but they should not be taken as mandatory nation- or region-wide staffing figures. On the other hand, however, the ratios between staff of different levels (e.g., first-line supervisors to field staff) can be derived from management experience and are relatively fixed. Another consideration should be that extension requires specialization and training to support and develop this specialization. Therefore, technical specialists must be technical specialists. Some staff members will always be better field officers than managers, but no one is a born manager, who does not require training in management.

Extension also requires a constant upgrading of the technical knowledge of staff. Appropriate training for each level and specialization of extension staff should be continuous. Regular training sessions to decide on and monitor field activities, usually given by senior extension specialists and researchers, should be complemented by specially designed and targeted training and study tours. A final consideration should be that the importance of an extension service should not be measured by its staff numbers or budget, but rather by the extent to which it attains its objectives at the least possible cost.

Salaries and incentives are an important issue. Absolute salary levels are important. But of equal concern to extension management is the trade off between the increasing ability of extension staff (and farmers) and the use of the media to support field extension activities, and thereby possibly replace some extension staff at the field levels initially and later at other levels, as well.

Two other points concerning salaries and incentives should be kept in mind. First, the prime venue of extension work is the farmers' fields. There are few other jobs for skilled staff that require such continual work in often difficult, unpleasant and isolated conditions. Carefully-targeted field allowances can help compensate for this. Second, as extension, more than many services, requires specialized skills in particular locations (a field staff and technical specialists, for example), promotion and pay structures should reflect this need. Specialists should be able to have a career as specialists and effective field workers should be able to remain in the field; both should receive recognition and awards in compensation for contributions to these areas.

Appropriate salaries and career structures make little sense if staff, once in position, are not able to work. Particularly where government employment has significant implicit welfare considerations, government priority is often on the payment of salaries and is in fact encouraged by budgetary processes that more readily accept recurrent rather than capital costs. In periods of resource constraints, salaries are the last element to be cut -- operational costs are usually the first to go, along with capital costs. For extension (and research and other services), this results in staff members not being able to do their assigned work. As a result, it has often been suggested that guidelines on the proportion of the extension budget that should be devoted to operational costs should be fixed (20-25% is the proportion commonly mentioned), and that the level of appropriate budgetary support for extension be set as a proportion of the agricultural domestic product, or some like measure. There is, of course, merit in such proposals, provided they are adhered to. Clearly, extension and farmers have an important function in advocating such guidelines. However, the reality of continuing resource constraints should be acknowledged, and ways for public extension services to work effectively within them identified. Some possibilities in this regard are taken up below.

Hardware

An Extension staff needs buildings, transport, training and audiovisual equipment and aids. However, while generalization is difficult, it is not uncommon for scarce resources to be used inappropriately on these necessary items. Duplicate or inappropriate training facilities may be built because of institutional rivalry. Priority may go to regional office buildings (which are rarely modest in size) when the need is for houses for field staff members. Moreover, houses may be built for field staff members when the incentives for them to rent are such that the houses remain unoccupied and gradually decay. Offices are provided for staff members when a greater priority is to keep them out of offices and in the field.

Transport is vital for a field extension service, but unless its use is supervised -- to be used by extension staff for extension work -- it can be a budgetary blackhole. Not discounting the numerous excellent examples of practical use, there are similar problems with audiovisual equipment. What proportion of still or video camera use is devoted to upgrading field activities and training, as opposed to recording the doings of dignitaries? As with all other budgetary aspects, extension management should continually evaluate the contribution of investment in hardware to attaining extension's objectives and adjust investment priorities accordingly.

THE TECHNOLOGICAL BASE OF EXTENSION

Effective agricultural extension must have a solid technological base. Extension's basic function is to advise farmers how to increase their production and income and, in light of direct frequent contact with farmers, to advise those who operate other agricultural support services how to improve their service to farmers. To do this, extension must have advice on crops and activities that is relevant to farmers. If such information is not available, there is little justification for significant expenditures on extension.

This does not mean, however, that there should be extension only where a significant stock of technology exists. There is sufficient variation in farming practices and standards in most farming communities, particularly in developing countries, for extension to work even when significantly improved technology is not available. Moreover, in such situations, extension has a major responsibility to ensure that research services, as well as other agricultural services, remain aware of farmers' problems and work towards relevant solutions. Just as extension cannot operate for long without strong agricultural research, neither can farmer-related research survive for long without extension.

While the need for a technological base for extension might seem a truism, it is frequently poorly handled in the design and operation of extension systems. It is often assumed that there is a backlog of messages that most farmers do not know or that they have not understood; that adoption of information is relatively simple and largely cost-free; and that extension's function is simply to diffuse messages. There is rarely a detailed analysis by extension (or in conjunction with research) of current farmer practices (particularly analyses specific to different resource conditions) and of the suitability of messages. Where such analyses have been done -- as in the agro-ecological zone agricultural technology status reports underway in India -- it has been shown that often few messages of immediate relevance to farmers are actually available, notwithstanding research and other claims to the contrary.

Numerous factors contribute to such elementary misconceptions. An important factor is often a cultural and perceptual gap that exists between extension (and research), on the one hand, and farmers, on the other. Education and class often divorce the extension agent from his clients, and particularly from poor small farmers. There are few research systems in which recognition and reward is accorded research results actually of use to farmers. Where extension (or any farmer support service) is not held accountable to farmers, there is a likelihood of such irrelevance.

Another factor that contributes to the on-going technological weakness of extension is a failure to observe and to listen. One example will suffice. How many field workers are convinced that the engineered soil and moisture conservation measures common to many countries are effective? Those who have seen the way they are breached by heavy rain, livestock and farmers and how they appear to contribute to water-logging and gullying certainly are not. However, rarely has such discomfort resulted in the development and promotion of alternative methods. Fortunately, these have now become available in the form of dry-stone walls and vegetative barriers.

A similar often uncorrected misconception due to weak observation by extension staff concerns the dry season. The absence of apparent farm activities in a marked dry season has led extension services and, therefore, other development departments, to believe that no extension activities need to take place. Indeed, in some countries, extension staff are not expected to begin training, let alone field activities, for the wet season until the rains have begun. Storage, processing and marketing of crops; ancillary farming activities, such as livestock management and soil and moisture conservation measures; selection and preservation of seed and purchase of inputs; and small-scale irrigated vegetable production are all examples of farmers' off-season activities to which extension can make important contributions. In addition, the extension staff should use this period to evaluate, in the field, their activities of the previous season, to follow-up with farmers or classes of farmers they did not serve well, to attend specialized training courses, to help analyze data from the farm trials in which they have been involved, and to plan priorities and strategies for the coming season. If time is still available, they can have special meetings with farmers to explain some of the basic scientific principles behind extension messages (for example, the functions of different micro-nutrients). Whatever time is left can be used for taking leaves.

A final factor I would like to single out as contributing to this failure to develop a strong relevant technological basis is the often poor development of adaptive research and the strict divisions that exist between extension and research. Priority in resource planning and resource allocation is often given to basic and applied research. Partly because responsibility for it is often diffused, adaptive research is frequently poorly managed and of little relevance to actual farm conditions. Especially where extension expects to receive technology from the black box of research, the consequence is that extension attempts to disseminate information that is irrelevant or uneconomic is excessively packaged (with respect to farmers' adoption and adaption processes) or is inadequately adapted to local circumstances. Fortunately, these problems are now widely recognized and a variety of mechanisms are used to overcome them. Among these, farming systems research has received perhaps the greatest publicity. Equally significant - for even farming systems research needs an extension service - are recognition of the dual functions of extension: (to guide farmers and farmer support services. The continuum of the agricultural knowledge system is also important, as is the more mundane creation of stronger extension-research linkages through mutual participation in strategy planning, joint field visits and field trials programmes, extension-research workshops to consider farmer feedback and to develop extension messages, and research participation in extension training (and, hopefully, one day, the reverse).

The commodity of extension is knowledge -- the knowledge of technology attractive to farmers and the knowledge of farmers' concerns and conditions for guidance to research and others. As such, training is a central element in any effective extension system. The field staff must be trained to listen and explain to farmers, as much as they are in technological developments. Supervisors need to be trained how to supervise, and technical specialists need training to attain and remain at the forefront of their specializations. Without training, an extension service cannot be sustained. Therefore, all possible avenues of training should be utilized within a comprehensive, detailed training plan. Opportunities occur locally for study tours, visits to research facilities, and short courses, as well as for selected academic upgrading. There is much to be learned abroad also, provided appropriate staff members are sent to relevant and challenging situations, and they are able to use the experience and knowledge upon their return.

THE SUSTAINABILITY OF EXTENSION

There are two main aspects of the sustainability of agricultural extension -- the technological and the financial. The foregoing section has made it clear that extension must have a technological base and that this depends in large measure on extension's orientation to farmers and the field, extension's influence on research, and the development within the service of an atmosphere of a need for knowledge. The more extension is able to respond to farmers' needs, the more it will be encouraged to concentrate on these, and so to be sustained technologically.

Financial sustainability is a different matter. Government-based extension services, our prime concern here, often have considerable recurrent costs in relation to government resources. While the rates of return on such investments can be established (Gershon et al. 1986), and cases may be made to attribute part of the change in the agricultural product to extension, the issue is usually not perceived in such terms. Government's chief concern, especially in circumstances where external financial support for extension is finite, is predictably more likely to restrain and even cut costs. Operational expenses, such as travel allowances, vehicle operation and maintenance costs, field trials and field activity recurrent costs, are usually the first to be affected; economizing on staff soon follows, often by assigning staff non-extension duties, even without transferring them from the extension service.

Within the extension service, a variety of steps can be taken to increase the financial sustainability of extension. One is recognition of the trade offs between field staff and the use of communications media. As the professional quality of the extension staff increases, and as the farming population's level of education rises, trade offs between quality and quantity of staff members are possible. Another important option is to take advantage of alternative extension options that may be available to farmers of particular crops and areas. These options are usually more applicable to large and more commercialized farmers than to poor small farmers, but they can be selectively used to enable the public extension service to limit its operations and to focus on more appropriate activities. For example, the public extension service may add to its functions the responsibility of ensuring that input dealers, cooperatives and even private voluntary organizations have access to up-to-date technical knowledge, and that its staff does not duplicate farmer contacts being made by such agencies. Elsewhere, the public extension service could provide technological backstopping to private extension agents, as has recently been begun in parts of India.

Major decisions concerning financial sustainability of extension also need to be taken outside the extension service. One basic decision is often that which concerns the extent to which parallel government extension services (and departments) can be merged. It is not uncommon in some areas for farmers to receive advice on production (as distinct from input support) from forestry, horticultural and livestock field staff, as well as from the agricultural service. Where production is primarily mixed, such field-level specialization is difficult to justify. Related to this is the need for all production encouragement strategies to be designed to take into account, wherever practical, the existing extension and input delivery systems. While this is obvious, not only can there be duplication of extension with previously established programmes, but there is often duplication with subsequently created and financed programmes. This issue is clearly one that a national policy framework for agricultural extension should tackle.

Another fundamental issue to be resolved outside the extension service is the extent to which farmers should be required to pay for extension support. This issue arises because of the cost involved in maintaining a public extension system, and because of the related awareness that the public sector is not always the most appropriate, and certainly not the only, supplier of extension services. A number of countries have systems of farmer payment for extension services (especially where a major commodity is concerned), but generally such payments are made indirectly.

There are circumstances in which farmers can be expected to pay directly and substantially for extension. Where production is focussed on a major commodity or where the general level of agricultural technology is well known, extension agents are likely to be most valuable when they provide a more individual advisory service by solving particular problems for particular farmers, and so their service is more suited to being individually billed. Most information on new technology, including some farmer-specific advice, is however more typically a public good -- that is, the provision of information to the user does not exclude others from using it. Indeed, because not all farmers can be provided with information individually, effective extension depends on the free dissemination of information among farmers.

All opportunities to recover extension costs from farmers should be utilized (as they should be for other services to farmers and other segments of the population), but direct recovery should focus on farmers and farming activities that receive individual advice with significant commercial implications. Where a market is developing for skilled and specific agricultural advice, government should reconsider its role in this market and evaluate its comparative advantage: it is normally sensible for a government to create conditions in which private suppliers of advice can emerge and flourish. While many extension needs of the poor, small farmers are not immediately suited to such advice, there is no reason why private-sector agencies should not be encouraged to serve such farmers because in the medium term they are likely to purchase other products purveyed by such parties.

BIBLIOGRAPHY

Baxter, M., R. Slade and J. Howell, "Form and Function in Agricultural Extension: Evidence from the World Bank and Other Donors". The World Bank, Washington, D.C.(in press), 1988

Gershon, F, L. Lau and R. Slade, "The Impact of Agricultural Extension". Staff Working Paper No. 576, The World Bank, Washington, D.C., 1986.

CHAPTER EIGHT

TECHNOLOGY UTILIZATION IN THE AGRICULTURE OF DEVELOPING COUNTRIES: ELEMENTS OF POVERTY-ORIENTED POLICY AND PROGRAMMES

Joachim von Braun 1/

THE NEED FOR EXPANDED TECHNOLOGY GENERATION AND UTILIZATION

There is a continuous need to rapidly expand the generation and utilization of new agricultural production technology in developing countries. I briefly point out the reasons for this need at an aggregate level, and then address a number of issues that are essential for the success of technology generation and utilization at the programme and farm level.

Ruttan (1975) conveniently separates material technology into, design, and capacity transfers. He points out that increased emphasis should be placed on capacity transfers for agricultural production technology in developing countries. The issue is to identify the appropriate mix of the types of technology generation or transfers for the specific country and local site. This requires substantial institutional development in which the regional, national and international agricultural research systems play an important role.

1/ Research Fellow and Coordinator for the Poverty Alleviation Policy Research Area , Food Consumption Program, International Food Policy Research Institute (IFPRI), 1776 Massachusetts Avenue N.W., Washington, D. C. 20036.

Trends in Production and Use of Food

Overall growth rates of food production in developing countries have slowed from the 1960s to the 1980s -- from 3.6 to 2.9% per annum. A rapidly increasing share of higher food production must come from incremental yields, rather than from extended areas, because areas for crop production are reaching their limits in most developing countries. The transition to this direction took place in North Africa, in the Middle East and in Latin America during the 1970s and is currently going on in Sub-Saharan Africa (Table 8.1). While overall yield increases accounted for about 80% of incremental food production during the 1970s, area expansion accounted for only 20%. Asia, especially, has limited room for area expansion as a means to increase food output.

The projected growth rate is increasingly dominated by expanded feed use. This growth rate is close to the one actually achieved during the 1970s when rapid technological change in agriculture, with large-scale adoption of new production techniques and inputs, took place. While potentially widening food deficits of developing countries could be filled in the short term with expanded food transfers from food-surplus producers in the industrialized world, the long-run solution lies in the further development and spread of new technology in the developing countries. Table 8.2 illustrates at the required growth rates of food availability to at least maintain the trend-projected developments.

Trend projections for developing countries, based on conservative assumptions of population and income growth, suggest a 2.7% per annum growth rate of domestic use of food and feed until the end of this century (Paulino 1986).

Poverty Orientation

Overall food availability is significantly important for nutritional improvement and poverty alleviation, although satisfactory aggregate supply alone does not assure adequate nutrition for everybody. Thus, there is a need to be concerned simultaneously with the rate of increase in food production and the means to increase production (Mellor and Johnston 1984). Unless a country's agricultural development facilitates the absorption of large increases in the rural labour force's productive employment, even a large increase in food output will leave many households with inadequate access to food supplies.

The efficient use of agricultural technology for poverty alleviation requires focus on incremental income of the poor. Such incremental income needs to be generated through expansion of employment -- farm and nonfarm -- in rural areas and through increased labour productivity in agriculture, in particular. Thus, technology generation that directly benefits the poor has to engender employment and simultaneously increase returns to labour.

In the following sections, some key issues are addressed for future technology development and technology utilization policies, especially at the programme and project level. Specifically addressed are programmes related to poverty-oriented approaches that promote technology utilization for small farmers. In this context, issues concerning adoption, profitability, sustainability, specific problems of women farmers, policy needs for support services (input supply, credit, human capital) and related constraints are addressed.

EXPANSION OF TECHNOLOGY UTILIZATION: AN AGGREGATE VIEW

Aggregate Developments: Seed, Fertilizer, Irrigation

The rapid expansion of new technology in food crops is indicated by the spread of new varieties, especially wheat and rice, in much of Asia and Latin America (Figs. 8.1 and 8.2). Although plant material has, and probably will, play an increasingly important role in the process of technological change in agriculture, it presents only a partial picture if the spread of new varieties is viewed in isolation. Plant material should, therefore, be examined along with the integral and complementary expansion of irrigation, fertilizer, extension services and plant protection. Developments in those fields, however, vary by region. Spread of new varieties and levels as well as growth of fertilizer use in Asia, for instance, exceed by far the developments in South America or Africa (Figure 8.3). Clearly, the limited scope for further expansion of production through increased crop area in Asia determined that pattern.

Irrigation played a crucial role in the rapid expansion of food production in developing countries. Nearly 40% of the world's foodgrains are currently produced on irrigated land. Recently, however, investment in irrigation has slowed in numerous countries as has the allocation of resources by some major donor agencies. This is partly due to reduced international foodgrain prices, to increasing costs per unit of irrigated land in marginal areas, and possibly to increased constraints on public resources for investment. IFPRI studies show that in Indonesia, for example, irrigation development was a major factor in the growth of rice production in the 1970s and 1980s. However, the rate of investment in irrigation decreased recently because of budgetary cutbacks, problems in the implementation of ongoing projects, and restructuring of priorities in the irrigation sector towards improved maintenance of existing systems (IFPRI 1987).

Maintaining and increasing yields in existing irrigation systems and diversification in the irrigation subsector are major challenges that lie ahead. Increased investment in institutions and human resources along with investment in plant material are required to meet the challenge. Although crop diversification may entail substantial potential, careful consideration of tradeoffs and risks is required. There may be high costs associated with a strategy of crop diversification that targets individual crops without considering their relative economic efficiency.

Complementary Factors

Various countries experiences differ in terms of technology utilization by crop and type of farmers. Interactions of factors such as technology generation, spread of technology, input supply policies, extension services, and institutional credit have become more complex in the agricultural growth process. A vivid example described by Ahmed (1988) is the case of Bangladesh during the 1970s and the 1980s when a number of factors interacted successfully for accelerated agricultural growth and development.

"A rice research institute was set up to organize research that meets various locational needs. Research scientists and agricultural extension officers were raised in status and salary to make them equal to engineers and medical doctors. This yielded very good results. Initially, IR-8 variety of rice spearheaded the expansion in areas under fertilizer-responsive, high-yielding varieties. Soon this was replaced by varieties of better adaptability in wheat research programmes. Research and extension service developed coordinated programmes for demonstration plots on farmers' fields. The progress on irrigation development, particularly privately-owned tubewells, made tremendous progress so that by 1985 about 20 percent of the rice area was covered by modern irrigation. A combination of irrigation and high-yielding varieties, supported by extension service and institutional credit, contributed to the rapid growth in fertilizer consumption during this period. Also during this period a T&V (training and visits) extension system was introduced in Bangladesh. The supply of institutional credit during this period was almost doubled in real terms compared to the supply during the 1960s. Irrigation and high-yielding varieties worked not only as direct accelerators in the use of fertilizer; but also, the indirect effect of these two factors was more powerful than the direct effect. Farmers who began to use fertilizer on irrigated high-yielding varieties soon spread the use to nonirrigated local varieties. To be specific, of the total consumption of fertilizers in 1984-85, about 60 percent was used on rainfed local varieties."

Production Variability and the Role of Policy

There is legitimate concern that increased use of the seed-fertilizer-irrigation technology in food production may, for various reasons, increase production variability, which then may impose burdens on the poor due to price and income fluctuations (Hazell 1986). However, fluctuations in production measured in terms of coefficients of variation of per capita food production increased, on the average, from 15.1 to 16.9%, comparing the periods 1961-74 and 1972-83 (the so-called post green revolution), while consumption fluctuations in per capita calorie consumption remained stable at 4.6 and 4.3%, respectively (Sahn and von Braun 1987).

It can also be shown that consumption fluctuations are much reduced where there is rising per capita income. Moreover, production fluctuations are much less translated into consumption fluctuations because of public policy on trade and stockholding, and because of the improved ability of households at higher income levels to adjust to supply and price changes (Sahn and von Braun 1987). The issue of production fluctuations as a potential consequence of technological change is addressed here to highlight the scope for development of public policy to parallel technological change, thus serving the purpose of alleviating poverty and improving consumption. A central role in this context is played by price policies as well as stockholding policies. A comprehensive assessment of the role of price policy in the context of new production technology is provided by Mellor and Ahmed (1988).

The core of technological change in agriculture is its reducing effect on per unit cost of production. In the context of cost-reducing technological change, prices can drift downward significantly without resulting in a disincentive to produce. Farmers not benefitting from improved technology may then shift to other commodities. "Thus, while the government may correctly choose to reduce price fluctuations, it needs sound analysis to distinguish that step from efforts to prevent longer-term changes in price relationships." (Mellor and Ahmed 1988).

Recognition of the complex relationships between technological change and price formation is critical for any poverty-oriented approach to technology development. Technological change tends to occur unevenly across geographical areas and is most likely to succeed where yields are already good. "Therefore, farmers in more prosperous regions are more likely to be favoured. Second, the disparities are likely to be further increased by strong multiplier effects of the new technology on the local economy. Third, if food prices decline as a result of increased production, regions not using the new technology will suffer because absolute as well as relative income declines." (Mellor and Ahmed 1988). These relationships, mediated through market or government interventions, or both, in the food and input markets, particularly shape the outcome of any attempted targeting of new production technology towards the poor -- those poor in developing regions, low-income small-scale farmers or landless labourers.

PRODUCTIVITY AND EMPLOYMENT FOCUS IN TECHNOLOGY GENERATION AND UTILIZATION

Direct and Indirect Effects on Poverty Alleviation

Poverty-oriented technology development aims at integration of the poor into a productive growth and development process. Directly, this can be achieved through development of technology and its dissemination into the control of poor small-scale farmers. Indirectly, the employment effects on landless labourers or small-scale farmers with labour surplus are of critical and growing importance. Finally, there could be potentially powerful multiplier effects emanating from increased rural income and employment (Mellor 1988). Increased spending on labour-intensive goods and services in rural areas would increase employment and income among the poor in periurban areas.

The direct effects of new technology on small farmers is important, but there are equally important indirect effects on the rural nonfarm population and on the increasingly sizable farm population that partly depends on off-farm employment. At programme and project levels, appropriate identification of the target group and its income sources and employment characteristics is critical.

Simplifications, such as "small-scale farmers equal the poor", or "ecologically disadvantaged regions with low output per unit of land are poor" are frequently used as targeting criteria. But these can be very misleading. Investment in socioeconomic research for appropriate target group orientation and technology development can have a high payoff in addressing the complex income and employment situation of the rural poor. These may, for instance, be concentrated in areas with high yields per unit of land on very small farms or constite the landless rural labouring class.

Technology Utilization and Crop Diversification

Increased real income and income stability are the two most powerful factors determining enhanced entitlements to food by the poor.

A recent study in The Gambia, West Africa, for instance, proves that in that area, incremental income from cash crop production of groundnuts and cotton, for example, similarly increases food energy consumption of the household, as does an equivalent increase in income resulting from incremental food production of rice (von Braun et al. 1987). This exemplifies two lessons: first, the most efficient mode of increasing real income of the poor through technological change can also be the most efficient way to increase food consumption, given the high preference of the poor to spend incremental real income on food. Second, it may well be that technological change in staple food production is an efficient way to achieve real income growth, and thereby nutritional improvement, among the poor.

Technological change in food crops has the incremental advantage to enhance local-level food supply, thereby increasing access to food, especially in backward regions. Due attention to regional market integration is required, especially when nonfood cash crops appear as the most efficient way to increase the real incomes of the poor, as local price inflation in food-deficit regions may be a consequence of a short-term drastic shift into cash crops.

A study of technological change in rice production in the North Arcot region of Tamil Nadu, India, based on data collected from household surveys in 1973-74 and 1983-84, showed dramatic increases in income and food consumption during this period. Daily calorie consumption per adult equivalent unit increased from 1,746 to 2,909, or an increase of 67%. The increase, however, was smallest for the landless, of which 10-15% continued to consume less than 80% of calorie requirements at the end of the study period. On average, it was estimated that about 38% of the income increase was due to increased rice production. This study also found that with the introduction of high-yielding varieties in the early 1970s, rice production increased by about 40%. Because input cost increased and the real price of paddy decreased during this study period, per hectare returns from paddy farming declined. However, paddy income was maintained by a combination of increased irrigation and short-season varieties that enabled farmers to grow more paddy, and also permitted farmers to expand their production of cash crops, such as groundnuts (IFPRI 1986).

Infrastructure and Technology Utilization

It is increasingly understood that the equity questions related to adoption or nonadoption of new technology by farmers is not properly addressed by merely comparing small-scale farmers and large-scale farmers. Adoption is, to a large extent, a function of the policy environment and the institutional supports -- input supply, credit, extension, marketing, etc. -- that bring new technology to farmers. The more complex technology development becomes, the more critical these factors are for the adoption rates. Programmes and policies, therefore, influence adoption. Adoption and technology utilization are not simply a function of farm and technology characteristics.

Technological utilization is not only a function of the nature of the production system, the nature of the technology, and the effectiveness of the institutions in transfering technology to farmers but also a function of rural infrastructure facilities. Research results from Bangladesh on this issue highlight that in villages with more developed infrastructure, the use of high-yielding varieties was 70% higher, fertilizer use 92% higher, and labour intensity in terms of days per hectare was 4% higher than in villages with less developed infrastructure (Ahmed 1988).

Rural infrastructure is also critical in the diversification of semi-subsistence agriculture. Income and employment gains from specialization are greatest for small-farm households, and are indirectly received by rural labourers who are close to well-functioning and maintained well infrastructure. This is exemplified in a study in Guatemala where it was found that villages further away from good roads ended up much less involved in income-increasing, labour-intensive, specialized vegetable production for export (Fig. 8.4).

Another case highlighting the importance of the interaction between technology utilization, market expansion, and rural infrastructure is identified in an ongoing IFPRI study in Zambia. Farmers who have adopted new technology through use of hybrid maize seeds, fertilizers, credit, and membership in extension programmes have higher income levels than nonadopters. Variations found among different villages within the same zone of the Eastern Province of Zambia point to the importance of infrastructure in institutional development (IFPRI 1987).

Technology Utilization for Household Food Security

Food security is the ability of households and their members to acquire sufficient amounts of food to lead an active and healthy life at all times. A complex set of interacting factors determines the effect of technological change on household-level food security. These are the effects of technological change on:

- employment and income in households that are food insecure;
- prices paid by these households;
- risks in production, producer prices, and employment; and
- intrahousehold income control.

Finally, to the extent that new technology changes the level and distribution of labour demand, the effects of this on "energy" expenditures (work load) may also play a role in food security.

An important aspect in this context is technology utilization that is not particularly risky for the small farmers. In an environment with risky food and labour markets (the latter is most important for the landless and small-scale farmers supplying a lot of labour to the off-farm labour market), crop production risk and price risk of the agricultural commodities become of central importance for adoption. Small-scale farmers will then shy away from high degrees of specialization -- be it in terms of variety selection within food crops or in terms of the switch from staple food production to nonfood cash crops for the market.

In the earlier example from Guatemala, for instance, it is found that more than 90% of the small-scale farmer households in the Western Highlands maintain a substantial share of their cropland under maize and bean production, although new export vegetables that are much more profitable and create substantial employment gains could be technically expanded. The main reason for this behaviour is that farmers consciously opt for a food insurance scheme that provides food from their own production. (The insurance premiums are the foregone gains from specialization.) The decision, however, is rational because of employment risk in the labour market and price risk in the food market and in the new export vegetable market.

Another interesting lesson learned from this example in Guatemala is that farmers who shifted substantial shares of their scarce land (average farm size 0.7 ha) into the production of export vegetables also increased their staple food output per unit of land. Yields of maize in these households increased, on average, by 30% because of the use of technology, especially more fertilizer, more labour for weeding, and, in particular, more timely crop management. Excluding fertilizer, which was already an accepted technology in the area, these were indigenous means by which food crop yields were increased jointly with the increased specialization towards export crops. Farmers consciously chose this for household-level food security. Technology development in such an environment can make a major contribution if it assists the small farmers in enhancing the staple food yields per unit of land (von Braun et al. 1989).

This example from Guatemala is not unique. The North Arcot study in India, referred to earlier, showed a similar tendency from a different angle: technological change in rice production had increased the yields in rice and jointly led to expansion of rice production with the expansion of cash-crop production (groundnuts). Moreover, the example from Zambia highlights that even those farmers with high rates of adoption of hybrid maize maintained a substantial share of their cropland sown with local maize, thus maintaining diversity in crop production within the same crop.

A general lesson from these examples of poverty-oriented technology development is that single-crop focus in technology development will only rarely match the farmer's objectives due to the associated risk of high degrees of specialization. Improving the whole income frontier of the rural poor and the small farmers must be the objective of technology -- be it in food crops, in nonfood cash crops, through direct or indirect employment generation via the forward and backward linkages in agricultural production, or through the potentially powerful multiplier effects for nonagricultural employment. To assess the country- and site-specific potentials and constraints for increasing income remains the challenge of successful programme design. The question "Who benefits from new technology?" must be asked continuously. It is not an issue of finding a final answer all at once because the conditions upon which the answers depend differ case-by-case and are changing -- that is, the characteristics of technology, the distribution of productive resources, the nature of institutional support, and local customs and traditions.

FOCUS ON WOMEN FARMERS

It is generally recognized that, from an efficiency and equity perspective, technology in agriculture must reach women farmers. The issue is of global importance and has gained increased attention as a result of attempts to stimulate agricultural production in Sub-Saharan Africa, where women play a dominant role in food production and marketing.

Translation of the important role of women farmers into policy and programme actions for technology development and diffusion leaves still much to be desired. Problems in implementing technology transfer mechanisms that actually reach out to women farmers suggest that the diffusion process itself requires a specific focus in research and programme development.

An example from The Gambia highlights an efficiency problem: labour productivity in women's fields, which in The Gambia can be clearly distinguished from men's fields, is substantially lower. This applies to basically all crops (Table 8.3). That this is largely a problem of technology utilization is highlighted by the fact that women farmers, who are the traditional rice growers in The Gambia, hardly gained access to the high-yielding crop technology disseminated in the study area (Table 8.4) (von Braun et al. 1988). While it is important from an equity perspective to observe the differential labour productivities between men and women farmers in this environment, the critical efficiency aspect of this assessment is that the difference in labour productivity between the women's and men's fields can be fully explained by the lack of access of women farmers to production technology (fertilizer, seed etc.) and to financial capital for more productive crop production (for hired labour, transport etc.). The latter highlights the eminent importance of credit to carry new production technology to cash-constrained women farmers.

Labour input in agricultural fieldwork varies by type of crop and farm size across regions of developing countries. In general, a high degree of substitution between women's share of household labour input in agriculture and hired labour input can be identified. With changes in agricultural technologies that lead to increased labour demand, it is generally found that household labour input, including women's labour, increases. Results from The Gambia, Zambia, and Guatemala indicate that improved technologies have such an effect on women's labour input. The situation, however, appears different in some cash crop-producing areas, such as those indicated in studies from the Philippines and Kenya, where an increase in sugarcane production was not associated with an increase in household or women's labour input. Hired labour filled the labour demand in both cases, as well as in the case for export vegetable production in Guatemala described earlier.

Such labour, production and employment effects of new technologies on women farmers are relevant in assessing the efficiency of a project. Lowering levels of women's labour productivity because of malfunction in technology delivery and input supply services is a waste of resources. However, there are also substantial socially negative equity and human welfare effects that result from a misallocation of public resources caused by the bypassing of women farmers on the part of technology systems.

IFPRI is accumulating evidence (from studies in The Gambia, Guatemala, and Kenya based on detailed household surveys) that enhanced women's income has incremental favourable effects on child welfare and nutritional improvements of vulnerable groups within households. To the extent that new production technology and its actual use lead to a relative shift of income control from women to male farmers, maximum benefits for child welfare and nutritional improvements are not obtained. In general, it is not the nature of technology per se to which women farmers have an unequal access, but it is in the institutional support services, especially input supply, credit delivery, and extension, which tend to bypass women farmers. Blaming maldistribution of potential benefits from technological change per se on technology is, in most instances, misguided.

TECHNOLOGY UTILIZATION IN HIGH-POTENTIAL AREAS VERSUS AREAS OF DIFFICULT ECOLOGIES UNDER POPULATION PRESSURE

Typical patterns of the effects of technological change in high-potential regions are depicted by the evolution in Punjab, India. Wheat yields rose from 1.1 t/ha in 1954-57 (pre-modern varieties period) to 2.7 t/ha in 1969-70. Total cost per hectare rose by 73%, yet because of higher proportional gains in output, cost per unit of output declined by about 30% (Ranade et al. 1988). In the 1970s and 1980s yield increases slowed considerably and the share of costs in total yield approached the pre-green revolution levels (Table 8.5). The cost structure changed, however, with substantially higher shares of costs for hired labour, machinery, and inputs (fertilizer, seeds, irrigation etc.).

A large body of literature argues that the new agricultural technology is a cause for rural inequality and poverty. Empirical evidence in general does not support this view. (See, for instance, Barker et al. (1985) and Hayami (1981) and the literature reviewed in those sources.) The outlook for the poor, however, appears disturbing, with a theoretical scenario of reduced technological change in agriculture: with rapid population growth, production pressure on land would increase rapidly without the absorption of an increased labour force into productive employment.

In rural economies of developing countries that experience a rapid increase in the man-land ratio, labour demand will not keep up with labour supply. Labour productivity, and thus wage rate, will deteriorate. Such reasoning is relevant for countries with large and rapidly increasing populations, with weak extension systems, and without substantial systems in place to generate the required growth in technological change. Rwanda may characterize such an environment: man-land ratios will rise from 5.5 to 12.0 adult person equivalents per hectare during 1985-2005 under cautious population and land use projections (von Braun et al. 1988).

Table 8.6 suggests, in a comparison of five survey sites, that returns to labour in food crops are lowest in the high altitude, high population pressure areas of Rwanda and Guatemala, compared to labour productivity in areas of some moderate technological change and under much less population pressure (Kenya, Philippines, The Gambia). There are good reasons to concentrate technology generation and extension promotion in areas of high population density, especially in Africa (Binswanger and Pingali 1988).

Technology development that permits resource preservation, sustainable agricultural production, and employment generation at increased labour productivity is called for in such regions -- as represented by the Central American or East African highlands. Employment needs to be directed towards capital formation in agriculture. Upgrading the agricultural resource base by erosion control measures can be a central activity in such high-altitude, mountainous ecologies. This justifies public investment (with temporary subsidy to labour) because of an inherent conflict between the need of the poor to cope with short-run survival and the long-run sustainability of the resource base for society's food security. Yet, the future for some of these ecologically volatile regions may be a mix of expansion in nonagricultural employment and outmigration into more productive regions where technological change would generate more rapid yet sustainable production and employment effects.

CONCLUSIONS

Population growth and decreased per capita income versus limited cropland in developing countries establish the increased need for expanded technology generation and adoption in agriculture. Increased food production alone, however, does not necessarily provide a solution to a poor household's food insecurity and malnutrition. New production technology, in order to benefit the poorest segments of the rural population, must raise labour productivity, increase employment, and thereby effectively raise demand for food. This means that to effectively alleviate poverty, the beneficial direct and indirect effects of new technology must reach rural households that have a limited land base and do not produce a net marketable surplus even under improved technology, as well as marginal production zones and the landless population. It is more the direct outreach via employment, also in the non-farm sectors (stimulated through growth in agriculture in high potential zones), that can reach out to the arid zones and the landless.

The process of technology generation and transfer will become increasingly complex. Improved education is of increased relevance because it enhances the potentials of success for technology utilization.

Investment in rural infrastructure provides the basis for the much expanded streams of goods, services, and production inputs required for accelerated technology utilization. Infrastructure improvement also facilitates diversification towards new crops (both food and nonfood) and the gains from regional specialization and interregional trade.

Public policy has to provide an environment conducive to technology utilization. Only then can well-designed research systems be effective in generating new technologies and extension systems.

Appropriate broad-based policy and technology generation vis-a-vis programmes for the spread and adoption of technology complement each other. All concerned, however, must not lose track of the specifics required to make technical change work in favour of the poor. Critical issues in this regard are the explicit concern for women farmers' labour productivity, the risk in prices and production under increased technology utilization, the profitability and sustainability of production under changed technology, and the constraints of small farmers in financing technology adoption. The latter requires increased attention to viable rural financial institutions catering to the poor. Of significant importance are the programmes and projects that support the buildup of local and national capacities in institutions that can effectively tackle these issues.

Policy to alleviate poverty in the rural economy of such countries and regions requires that:

- the new technologies for crops and production systems that are introduced in most environments must focus on increased labour productivity and expanded employment;
- new crops and technologies must take into account the complex patterns of seasonality in production and processing, both of which impinge upon the opportunity cost of labour, and require investment in locally applied agricultural research capacities;
- the introduction of new technologies must not require large amounts of working capital on the part of the farmer until working credit systems are set up. This is especially true for women farmers who lack access to credit to finance technological change;
- the marketing channels for inputs and outputs of newly introduced crops must not be excessively risk-prone; there is a key role for credit and savings assistance in coping with risks;
- new technologies and crops must take into account the agricultural resource base and environment and assure that soil fertility and sustainability of subsistence agriculture is not at risk;
- production, consumption, and health are inseparably tied together in the rural economy and have to move together. Rural infrastructure and services play a critical role in this respect.

Table 8.1

AVERAGE ANNUAL GROWTH RATES OF PRODUCTION OF MAJOR FOOD CROPS AND THE CONTRIBUTION OF AREA AND YIELD TO PRODUCTION INCREASE, 1961-70 AND 1971-80

	Growth rate per annum (%)		Contribution (%) to production increase from			
	Production		Area *		Yield/ha	
	1961-70	1971-80	1961-70	1971-80	1961-70	1971-80
Developing countries	3.6	2.9	30	20	70	80
Asia	3.8	3.3	13	16	87	84
N. Africa/Middle East	2.4	2.6	51	26	49	74
Sub-Saharan Africa	2.2	1.6	100	50	– **	50
Latin America	4.2	1.8	66	33	34	67

* Negative rates also occurred for these areas during subperiods.
** The contribution to the production increase was assigned totally to the other source of increase.
Source: Paulino 1986, p. 22.

Table 8.2

PROJECTED GROWTH RATES OF DOMESTIC USE OF MAJOR FOOD CROPS, 1980-2000

	Growth rates per annum (%) *
Developing countries	2.7
Asia	2.3
N. Africa/Middle East	3.8
Sub-Saharan Africa	3.6
Latin America	3.2

* The growth rate of domestic use is based on 1980 trend estimates and projections to 2000 on trend income growth.
Source: Paulino 1986

Table 8.3

GROSS MARGINS (NET RETURNS) PER PERSON PER DAY IN WOMEN'S AND MEN'S FIELDS, THE GAMBIA (WET SEASON 1985)

	Average gross margin per person per day*		
	Women's fields	Men's fields	
	(US $)	(US $)	(in % of women's)
Full water control (new scheme)	-- **	2.32	--
Part water control (new scheme)	1.30	-- **	--
Traditional rice	0.90	1.48	164
All rice***	1.01	2.09	206
Millet, sorghum	1.33	1.52	114
Groundnuts	1.08	1.67	155
Cotton	0.33	0.65	197

* Converted at parallel exchange rate (US$1 = Dalasi 6)

** Less than 25 observations.

*** Area weighted average (fields under each group's control)

Source: IFPRI/PPMU survey, 1985-86

Table 8.4

WOMEN'S RESPONSIBILITY FOR RICE FIELDS AND TECHNOLOGY YIELD LEVELS, 1984

	Average yield (t/ha)	Fields under women's responsibility (%)	Fields under cash crop (%)	Family labour done by women (%)
Fields with full water control, new rice scheme	6.6	14	99	29
Fields with pump irrigation, old rice scheme	2.9	54	67	68
Fields with part water control, new rice scheme	2.2	70	87	60
Traditional swamp rice	1.3	95	17	77

Source: IFPRI/PPMU survey 1985-86

Table 8.5

COSTS AND RETURNS OF WHEAT PRODUCTION IN PUNJAB, INDIA

	1954-57	1969-70	1978-79
Yield (t/ha)	1.1	2.7	2.7
Total cost (% share in yields)	66.9	48.0	64.6
Share in total costs (=100.0)			
Hired labour:	8.0	19.8	15.1
Family labour	28.1	15.4	6.6
Bullock labour	37.9	15.7	6.9
Machine labour	--	7.4	15.4
Fertilizer, seed, irrigation	15.5	31.3	36.4
Other	10.5	10.4	19.6

Source: Ranade et al. 1988

Table 8.6

RETURNS ON LABOUR IN STAPLE FOOD PRODUCTION IN VARIOUS SETTINGS IN THE MID-1980s

Survey*	Crop	Net return (US $) per/family labour day**
Philippines	Maize	2.15
Guatemala	Maize	0.70
Kenya	Maize	1.45
Rwanda	Maize	0.49
The Gambia	Millet	1.36

* Surveys are not representative of entire respective countries but of regions of them only.

** National currency converted at parallel exchange rates to account for over-valuation.

Source: Various IFPRI surveys, 1984-85

Figure 8.1

ESTIMATED PROPORTION OF AREAS PLANTED WITH HIGH-YIELDING VARIETIES OF RICE AND WHEAT, SOUTH AND SOUTHEAST ASIAN COUNTRIES, 1965-66 to 1982-83

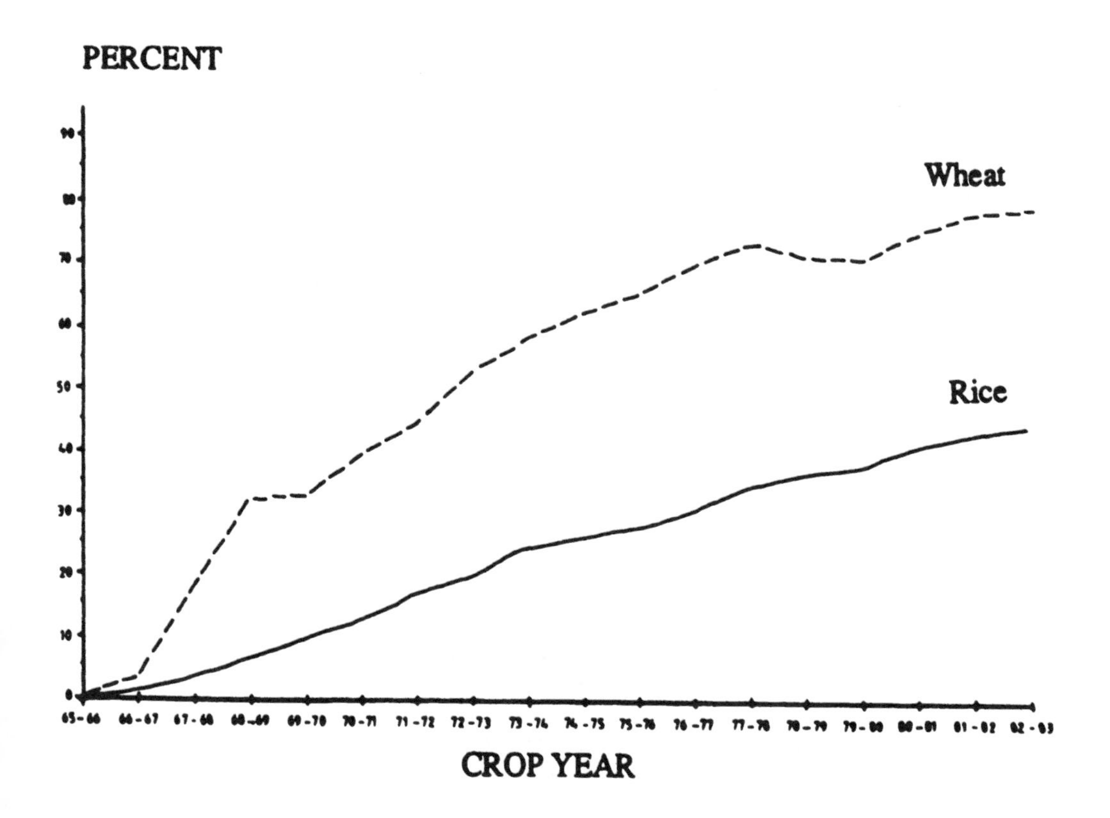

Source: D. Dalrymple, Development and Spread of High- Yielding Rice Varieties in Developing Countries (Washington, D.C.: U.S. Agency for International Development, Bureau for Science and Technology, 1986), p.110.

Figure 8.2

ESTIMATED PROPORTION OF TOTAL AREAS PLANTED WITH HIGH-YIELDING VARIETIES OF RICE, 22 LATIN AMERICAN COUNTRIES, 1969-70 TO 1981-82

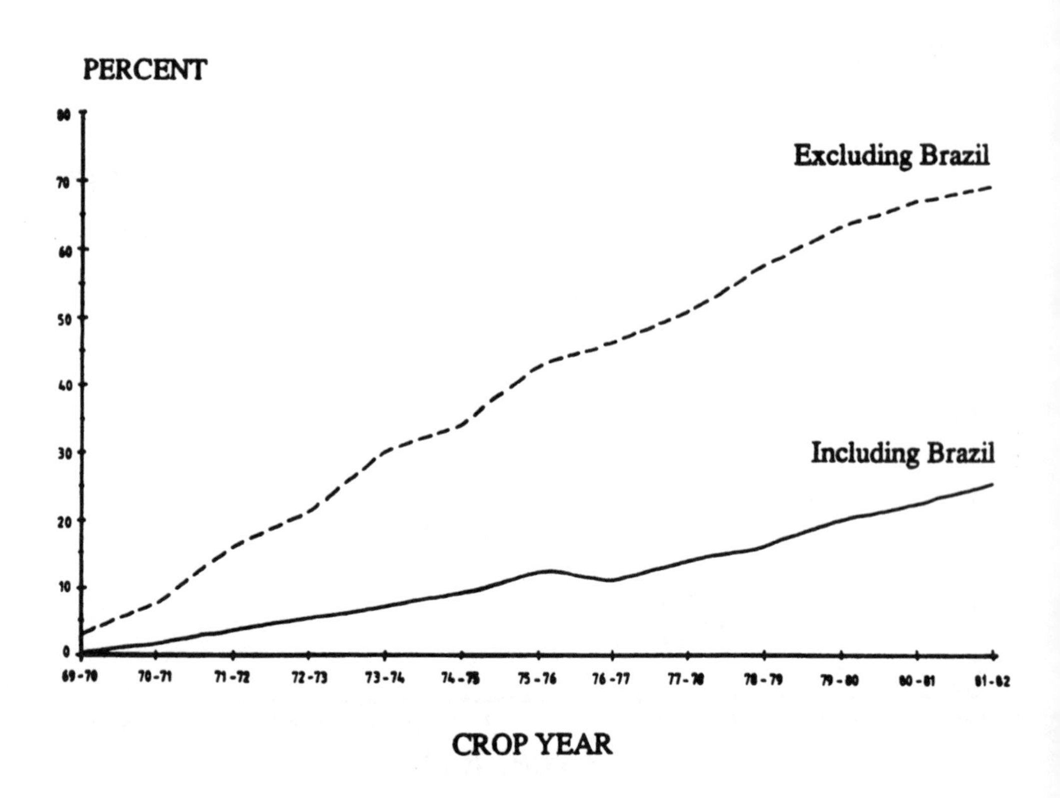

Source: D. Dalrymple, Development and Spread of High -Yielding Rice Varieties in Developing Countries (Washington, D.C.: U.S. Agency for International Development, Bureau for Science and Technology, 1986), p.110.

Figure 8.3

FERTILIZER CONSUMPTION PER CULTIVATED LAND AREA
(1968-1984)

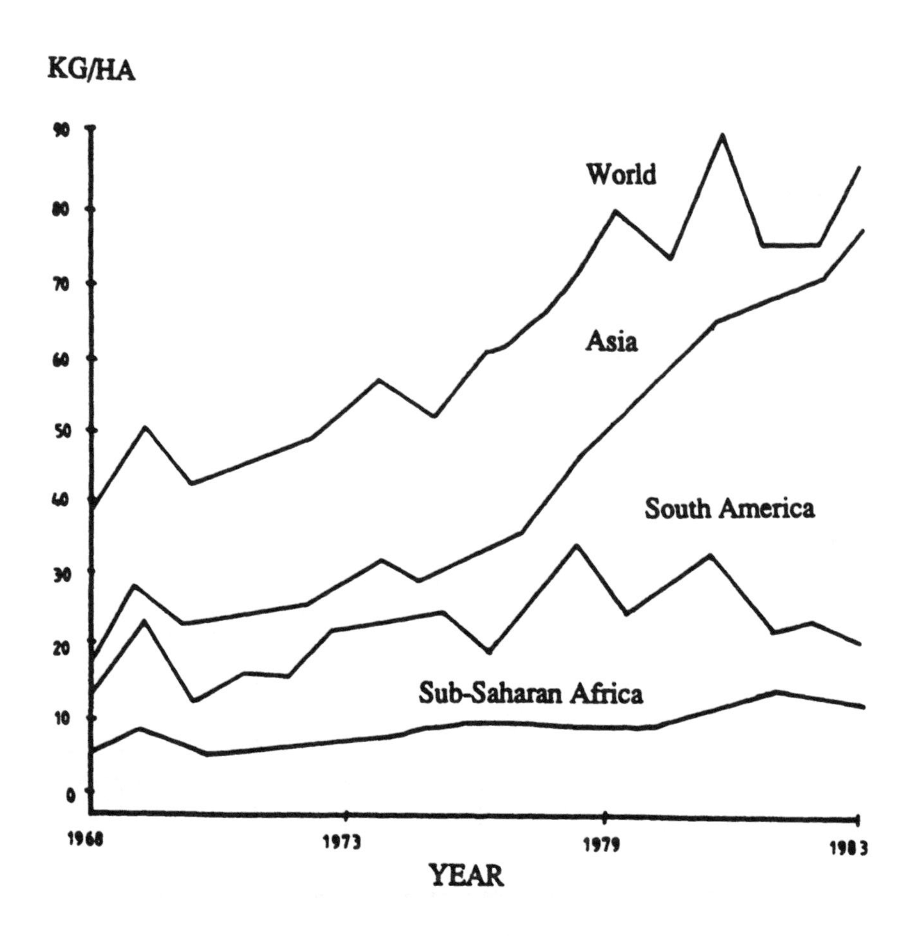

Source: IFPRI Annual Report, 1986, based on Food and Agriculture Organization, Fertilizer and Production Yearbook (Rome: FAO, various years).

Figure 8.4

PROXIMITY TO PAVED ROAD AND SHARE OF LAND USE FOR CASH CROPS IN SIX VILLAGES IN GUATEMALA

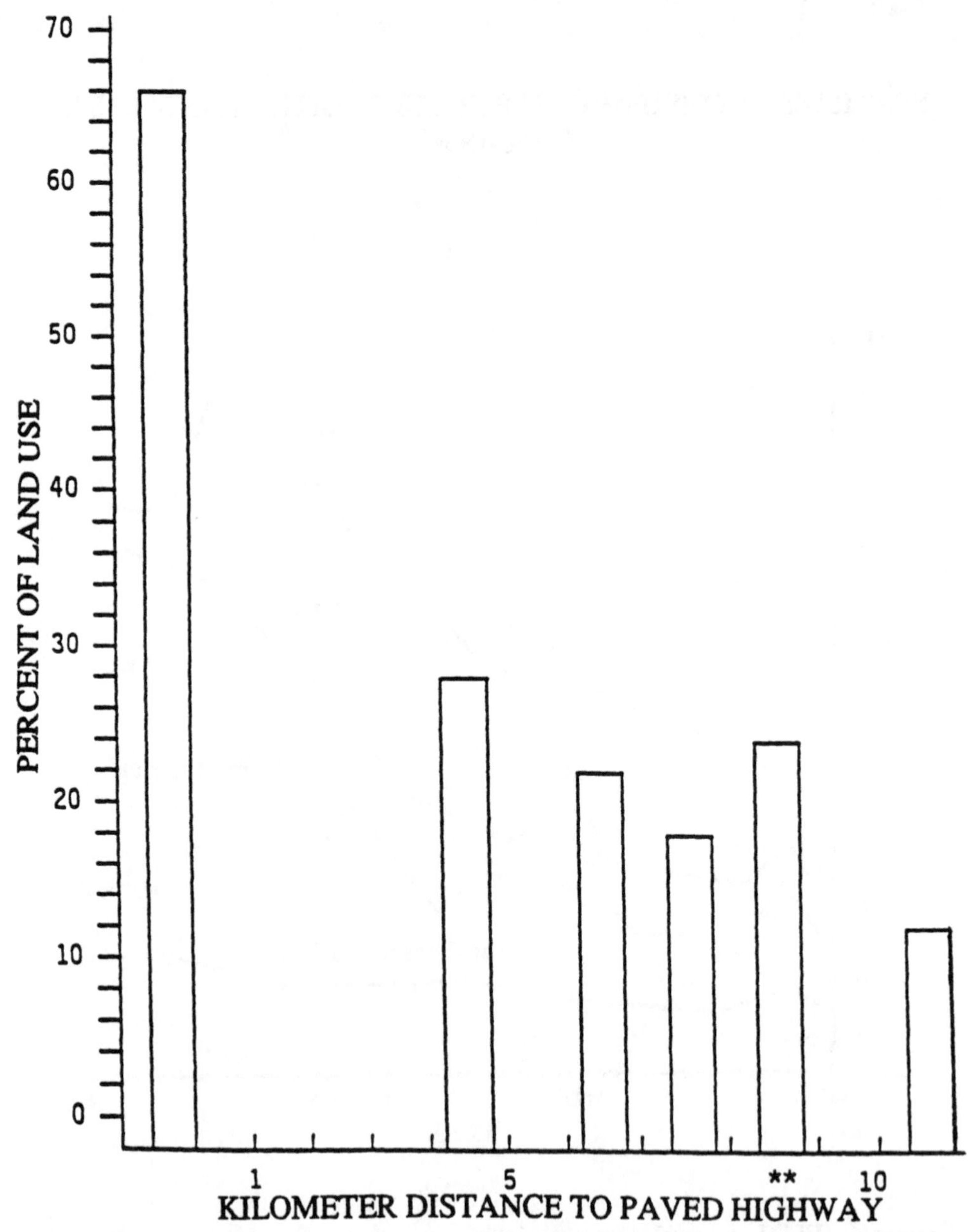

** This village was an early member of the export vegetable cooperative.

Source: IFPRI/INCAP survey, 1985.

BIBLIOGRAPHY

Ahmad, R., "Agricultural Production, Employment and Income". "Infrastructure and Agricultural Development". IFPRI policy briefs No. 3. International Food Policy Research Institute, Washington, D.C., 1988.

Ahmad, R., "Fertilizer Use in Asia: Lessons from Selected Country Experiences". Paper presented at the IFPRI/IFDC workshop on Fertilizer Policy in Tropical Africa, Lome, Togo, 1988.

Barker, R., R.W. Herdt and B. Rose., The Rice Economy of Asia. Resources for the Future, Washington, D.C., 1985.

Binswanger, H. and P.Pingali., "Technological Priorities for Farming in Sub-Saharan Africa. Research Observer 3 (1) : 81-97, 1988.

Dalrymple, D., Development and Spread of High-yielding Rice Varieties in Developing Countries. U.S. Agency for International Development, Washington, D.C., 1986.

Hayami, Y., "Induced Innovation, Green Revolution and Income Distribution: A Comment." Economic Development and Cultural Change, No. 1, Oct. 1981.

Hazell, P. (ed.), Summary Proceedings of a Workshop by IFPRI and DSE on Cereal Yield Variability. International Food Policy Research Institute, Washington, D.C., 1986.

International Food Policy Research Institute. 1987 Annual Report. Washington, D.C.

International Food Policy Research Institute. 1986 Annual Report. Washington, D.C.

Mellor, J.W., "A Food and Employment-oriented Development Strategy: Needs, Problems and Opportunities in the Emerging World Food Situation and Challenges for Development Policy". IFPRI Policy Briefs, No. 11, International Food Policy Research Institute, Washington, D.C., 1988.

Mellor, J.W. and R. Ahmed eds., Agricultural Price Policy for Developing Countries, Johns Hopkins University Press, Baltimore, 1988.

Mellor, J.W., and B.S. Johnston, "The World Food Equation: Interrelations Among Development, Employment, and Food Consumption", Journal of Economic Literature 1984, 22: 532.

Paulino, L.A., "Food in the Third World: Past Trends and Projections to 2000. Research Report 52, International Food Policy Research Institute, Washington, D.C., 1986.

Ranade, C.G., D. Jha and C. Delgado, Technological Change, Production Costs, and Supply Response, J.W. Mellor and R. Ahmed, eds., Agricultural Price Policy for Developing Countries, Johns Hopkins University Press, Baltimore, 1988.

Ruttan, V., Technology Transfer, Institutional Transfer, and Induced Technical and Institutional Change in Agricultural Development, L.G. Reynolds,ed., Agriculture in Development Theory, Yale University Press, New Haven, 1975.

Sahn, David and J. von Braun, "The Relationship Between Food Production and Consumption Variability: Policy Implications for Developing Countries", Journal of Agricultural Economics 38 (2): 315-327, 1987.

von Braun, J., H. de Haen and J. Blanken., "Process and Effects of Commercialization of African Agriculture in a Most Densely Populated Area (Rwanda): Consequences for Food Security Policy". Report submitted to the German Technical Assistance Organization. International Food Policy Research Institute, Washington, D.C., 1988.

von Braun, J., D. Puetz and P. Webb., "Technological Change in Rice and Commercialization of Agriculture in a West African Setting: Effects on Production, Consumption, and Nutrition". Report to the International Fund for Agricultural Development, International Food Policy Research Institute, Washington, D.C., 1987.

von Braun, J., D. Hotchkiss and M. Immink., "Nontraditional Export Crops in Traditional Smallholder Agriculture: Effects on Production, Consumption and Nutrition". Research report (in press). International Food Policy Research Institute, Washington, D.C., 1989.

CHAPTER NINE

EXTENSION SUPPORT ACTIVITIES AND TECHNOLOGY UTILIZATION FOR POOR SMALL FARMERS

G. Edward Schuh 1/

I have seen extension from a number of different perspectives -- as an extension worker and administrator and as an advisor on funding extension programmes in developing countries. I have also done considerable research on extension and related issues. Some of my research focussed on evaluating the effectiveness of extension services. Other research focussed on factors that influence the adoption of technology by small farmers and on understanding the economic performance of small farmers.

My experience causes me to take a rather broad approach to the issues before us. It also gives me what many would view as an unconventional approach to the problem. I have decided to approach the problem from the broad perspective, largely because I believe such an approach is consistent with my assignment, and because I believe we need to consider these broader issues.

I divide my remarks into two unequal parts. The smaller part focusses on some issues raised in the agenda for this seminar. I state my views on some of the issues raised with little defense or

1/ Dean and Professor, The Humphrey Institute of Public Affairs, University of Minnesota, Minneapolis, Minnesota.

explanation. In the remainder of my remarks I focus on a broader set of issues that I believe are critical in developing effective extension services.

SOME GENERAL PROPOSITIONS ABOUT EXTENSION

Briefly and to the Point on Issues:

1. In general, extension services do not make sufficient use of the communications media to attain their objectives. Moreover, among the various media, they tend to under-utilize radio. The more widespread use of magazines and newspapers, unfortunately, provides a bias to the middle-sized and large farmers. Small farmers are typically illiterate and not able to take advantage of the printed word.
2. Monitoring and evaluation are important instruments for institutional development. Although often used as a single concept, I believe there are two distinct issues here and the procedures for each need to be different. I tend to view monitoring as a management information system, which enables administrators to better understand how their programmes are doing. Evaluation should be a more in-depth analysis of the impact of particular progammes, of their effectiveness, and of what is needed to improve them. The focal points for both operations should include what happens at the farmer level and what is happening within the extension organization per se.
3. Raising the income of farm families should be the ultimate objective of agricultural extension. That is a powerful proposition with powerful implications. It is also likely to be controversial. Yet to take anything else as the objective will lead to contradictions and loss of effectiveness in extensions programmes.
4. In general, senior and middle level officers do not, in my judgement, get out to the fields often enough. In some cases, not even lower level staff members get out often enough.
5. The best incentive system for an extension service is a differential and merit-based salary structure. People should be rewarded for good performance. Poor performers should not be rewarded purely on the basis of longevity in the service.
6. Certain physical inputs are highly complementary to new knowledge, especially when that new knowledge is not imbedded in physical inputs themselves. Thus, the availability of these complementary inputs, such as the fertilizer and water that need to go with improved varieties, or knowledge about improved spacing and timing of treatment, should receive high priority. However, it is usually counter-productive for extension people to get involved in the provision

of these inputs although often they have an important role to play in assuring that the necessary inputs are made available by other agencies or entities.

SOME BROADER ISSUES

There are broader issues that implicitly bear on the issue of support activities and technology utilization. These include:

1. skills development versus message transfer;
2. the significance of economic policy and principles;
3. the need for transformation within agriculture;
4. community development; and
5. the marginal lands - sustainability issue.

Skills Development Versus Message Transfer

The World Bank's emphasis on the training and visit (T&V) extension system has wrought a veritable transformation in the approach to extension worldwide. Although I find much to admire and respect in this system, I also have some serious reservations about it as a general approach to extension, and, in some cases, I believe it can take us in the wrong direction in developing viable extension services.

My general position on the T&V system is that it can be effective as a means of revitalizing a moribund extension service that has been in place for some time. That tends to be where it has had its greatest success. However, if one were be designing a new extension service, I suspect there is little from the T&V system one would want to adopt -- largely because it is too labour-intensive and expensive and because it inherently, adopts a wrong approach.

What should an extension service do? Ultimately, its objective must be to develop the skills and abilities of farmers and rural people. What does the T&V system do? It tends to focus on identifying useful knowledge messages and diffusing them among farmers and rural people. There is little development of cognitive skills or of the ability to think things through. In fact, the T&V system often reminds me of that classic description of the U.S. Navy from The Caine Mutiny: "A system designed by geniuses to be implemented by fools!"

I believe that any extension system that is designed to have a significant impact over time must take its mission to be largely educational. It should be teaching principles and how to use those principles. Knowledge has to be adapted to the individual circumstances, and farmers themselves are best able to do that. Producers and others need to develop their own skills and their own initiatives. Spoon-feeding them with messages will do neither.

The emphasis on education implies a great deal about how an extension staff should be organized, about how it should be trained and developed, and about what the supporting services and systems should be. Obviously, we are not talking about an information or message transfer system. We are talking about a knowledge system in which the knowledge is expressed in principles and not bits of information. Moreover, rather than patronize the producer and assume that he or she cannot learn and adapt the knowledge, it attempts to empower them with new knowledge on as broad a scale as possible -- perhaps the most liberating thing that can be done for them.

The Significance of Economic Policy and Principles

In my experience there has been nothing more frustrating than to see extension agents making recommendations or delivering messages that are completely contrary to prevailing economic incentives or conditions. In one extreme case I observed many years ago, the extension service in the state of Minas Gerais, Brazil advised farmers to apply fertilizer to their maize crop when all the evidence suggested this was not a profitable practice. At the other extreme, I recently observed members of a T&V team in Kenya recommending that farmers adopt a double-row system of maize and beans. This is a labour-intensive system and wages were high in the region. Farmers were smart enough not to adopt the system. The managers of the T&V team thought the staff were not doing their jobs!

The fundamental problem with these uninformed and misguided recommendations is that they cause the extension service to become discredited. The farmers knew the recommendations were not profitable. Or if they didn't, they soon learned. And once they knew or learned, their only conclusion could be that the extension staff didn't know what they were doing -- or what they were recommending. How much better it would have been to either make a careful economic analysis prior to making the recommendation -- an important support service -- or to have taught the farmer the economic principles involved in fertilizer use, (or the double-row system), and thus enable him or her to make their own decision.

The need to strengthen the economic core of our extension training is a point that I want to make in this section. This should be a comprehensive programme that includes principles of marketing as well as principles of production. The second point I want to make is the need to provide economic information in the form of outlook and analysis of contemporary developments as part of the support services to extension programmes. Unfortunately, all too often we become so wrapped up in the technology side that we forget the inherent complementary nature of technology and economic information, and the need for both kinds of information.

There is a third issue: the role of economic policy in making new production technology profitable and the role of the extension service in helping to obtain improved policy. It is no secret that economic policy in most developing countries discriminates against agriculture by shifting the domestic terms of trade against it. This shifting of domestic price relatives often makes it unprofitable to adopt new production technology. The case of fertilizer is classic in this regard. And if the use of fertilizer is not profitable, it generally is also not profitable to adopt more fertilizer-responsive varieties.

There are other distortions to domestic price relatives that are equally as important. For example, highly subsidized credit makes it rational for the individual farmer to mechanize, whereas socially this may be wasteful of economic resources. No amount of telling farmers not to mechanize under these circumstances will do. What is needed is to change the policies.

At still another level, tax policies that discriminate against agriculture often have a differential effect by size of farm. Large farms can escape such taxes by reorganizing their resources to produce on a more extensive scale. This means they have little incentive to adopt new technology. The small producer cannot escape such taxation, and thus bears the major burden of the tax. Often they will, under these conditions, have incentive to use the new technology, but they will be subject to such severe capital rationing -- internally and externally -- that they are unable to adopt the technology.

The sum and substance of this situation is that the social payoff to investments in the extension service may be very low -- even though the service may be effectively organized and capable of delivering knowledge and education on an extensive scale. Under these conditions, neither society nor policy makers will be interested in supporting the extension service. Hence, it is an imperative of viable extension services that they give greater attention to economic policy and economic principles. Extension services, perhaps more than any other part of the agricultural infrastructure, can make the case for improved economic policies.

The Need for Transformation Within Agriculture

I am struck by the tendency we often have to take agriculture as a static receptacle into which we pour new production technology as a new social input. Such a perspective has dominated much of the discussion at this seminar. Farmers are viewed as static units to which new technology is to be transferred, with the result that they will all be better off.

That's not the way it works, however. New production technology typically has a differential effect in enhancing resource productivity. Mechanization, for example, typically raises labour productivity with little effect on the productivity of land. Biological technology, on the other hand, tends to raise land productivity and only under certain circumstances to raise the productivity of labour. Thus, depending on the character of the technology, there can be significant effects on resource proportions. Moreover, certain forms of production technology, such as mechanization, can create great pressure for farm enlargement, and with it the discharge of workers from agriculture.

Still another factor at work is the forces of economic development. As per capita incomes in the nonfarm sector rise, farm size in the agricultural sector has to be increased if farmers are to receive incomes comparable to those in the nonfarm sector. This is a powerful force for farm enlargement and the exodus of labour from agriculture. Yet how often do we see recognition of this in our extension programmes? How often do we see extension services providing training for off-farm employment? Yet that is a clear implication of these powerful economic forces.

Community Development

This need to transfer labour out of agriculture as development proceeds has another important implication for extension. The need for this transfer is one of the reasons that per capita incomes in agriculture typically lag behind those in the nonfarm sector. The problem is exacerbated by the fact that seeking employment in the nonfarm sector typically involves a costly geographic move. This causes the wage and income differentials to be even larger.

This dependence on a neoclassical migration model as the basis for bringing about equity among sectors is counter-productive. It tends to extract the best human capital from the low-income agricultural sector and to give it as a donation to the upper-income urban sector. A more rational process would be to decentralize the development process, in effect taking the new employment opportunities to where the labour is, rather than the other way around. In this way the human capital remains, and is, the basis for further development.

The implication of this for the extension service is that soon after the modernization process gets under way in agriculture, high priority needs to be given to rural and community development programmes. This is not the place to go into detail on these programmes. But the most effective way of attracting nonfarm activities to the rural sector is to assure that an adequate physical and social infrastructure is in place and that the labour force has the skills for employment in the non-farm sector. Such rural and community development activities are critical support activities to technology transfer programmes. We need to get them higher on our policy and development agendas.

The Marginal Lands/Sustainability Issues

The issue of marginal lands and marginal workers, together with sustainability, has recently risen high on the agenda in agricultural development circles. I must confess to having more than a few concerns about the present level of debate on these issues.

First, I believe there is a confusion between small-scale producers and producers on marginal lands. There is ample evidence that focussing development activities on small-scale producers in general has a high social payoff. That is a rather different proposition than whether focussing development activities on marginal or inherently unproductive lands has a social payoff. My concern, in particular, is that pleas are frequently made to shift scarce research funds to raising productivity in these inherently poor regions, and away from productive regions where the payoff may be much higher.

Determining whether land is inherently poor is not an easy matter. However, we need to pose the question and try to determine where payoffs from the application of scarce research resources will be the highest. In those areas in which the potential for productivity payoffs is limited, we need to develop programmes that attempt to shift labour to nonfarm employment. We are thus back to the need for rural and community development programmes.

Second, there seems to be a similar confusion about sustainability issues. In many cases, the costs of obtaining a sustainable agriculture may be quite high. This implies that the social rate of return to such investments may be quite low. If that is the case, the effort should be directed to inducing the population off the marginal lands and into alternative employment, rather than to attempt to sustain systems that are inherently not sustainable.

CONCLUDING COMMENTS

Many of my remarks point to the need for a stronger economic component to our extension programmes, and to the need for a more careful economic analysis of what we try to do with our extension programmes. Our general failure to take account of this important dimension often leads to low productivity for the system, discrediting of extension services, and to serious misallocation of scarce development resources. This suggests that a strong capability for economic and social research within the extension service can potentially have a very high social payoff. This capability may be the most important support activity we can provide to our technology transfer programmes.

CHAPTER TEN

TECHNOLOGY SYSTEMS FOR RESOURCE-POOR FARMERS

Abbas M. Kesseba 1/

INTRODUCTION

This paper attempts to draw attention to the major issues highlighted at the IFAD Seminar on "Generation and Transfer of Technology for Poor Small Farmers" held with the assistance of the National Agricultural Cooperative Federation (NACF) in Seoul, South Korea, in June of 1988. The Seminar drew representatives from several developing countries, bilateral donor agencies, multilateral financing institutions, international institutes and UN agencies and was held with the objective of seeking a better understanding of the issues and options related to technology generation and its transfer to poor small farmers.

The theme of the Seminar was agricultural research and extension, which has been a subject of many workshops and conferences. However, this Seminar was the first attempt to discuss the subject solely from the perspective of the poor smallholder farm. The synthesis below underscores this need, especially when it is generally recognised that in most developing countries the small-farm sector is often neglected by Technology Systems in favour of the large-farm sector.

1/ Coordinator, Technical Advisory Unit, International Fund for Agricultural Development, Rome.

The contents of the following sections discuss the major issues relating to (i) The small farm sector and the focus of research and extension, (ii) research-extension linkage and the small farmer, (iii) investments in agricultural research and training, (iv) the relevance of agricultural research for small farmers, (v) extension systems, (vi) extension support, and (vii) technology utilization for sustainable agricultural development.

THE SMALL FARM SECTOR AND THE FOCUS OF RESEARCH AND EXTENSION

The small farm system is characterized by a high degree of complexity and diversity. This is expecially true with regard to the small farmer's aversion to risk, which dictates that the production process always includes both mixed farming and multiple cropping. As a result, the production system involves the intensive use of the meagre resources of land and family labour and minimal use of inputs external to the farm. The smallholder's survival is based on a production system which is susceptible to stress factors, including demographic pressure. Moreover, at the subsistence level, the nominal income thus generated is highly variable from season to season. These variables make the small farming system unstable and eventually lead to its disintegration.

Small farmers often form the largest group in the agricultural sector. Their traditional technology systems, which have evolved over the centuries, are basically sound ones and are adapted to cultural and environmental conditions. The traditional skills and methods used are often less expensive and technologically more appropriate than those imposed by development agencies in general. However, the traditional production systems of smallholders are being increasingly polarised towards marginal agro/ecosystems in many marginal areas; demographic pressures and increased land fragmentation are causing smallholders to intensify resource use, which progressively depletes their resource base. These factors eventually result in the breakdown of traditional systems.

Research and Extension (R&E) should play a very crucial role in the small-farm sector. However, R&E has generally tended to overlook the small farm because the entire Technology System in most developing countries is not geared to it. In order to address small-farmer issues effectively, R&E needs to consider the unique constraints associated with small farming systems. Focus on traditional crops and on the farming system as a whole is essential if R&E is to become relevant to smallholders. Recognition of the close association of widespread hunger and malnutrition with small subsistence farming systems should be reflected in R&E activities addressing smallholders and the substantial untapped production potential they offer.

Small farmers represent a wide range of different categories including marginal, subsistence farmers and farmers who, in spite of their limited resources, may produce marketable surpluses. Specific research is needed to meet the diverse needs of these small farmers. The farmers' aversion to risk and their limited resource base should be important considerations for designing R&E interventions in the small-farm sub-sector. R&E should also take into cognisance the inevitability of resource-poor farmers' intensive interaction with the environment and its resultant degradation.

Furthermore, R&E must also recognize the critical role of women, both as small farmers and as important members of the rural small-holder household. Therefore, in order to address poor smallholder farmers or rural poverty meaningfully, the focus of R&E should not be merely on factors associated with farm production but on the wider implications of R&E in terms of overall (intra) household welfare.

RESEARCH-EXTENSION LINKAGE AND THE SMALL FARMER

A number of important issues still have to be resolved for the generation and transfer of technology relevant to small farmers. The linkage between research and extension is a key issue. Historically speaking, they have developed independently and still tend to work separately. In developing countries, in recent decades, R&E linkages have had a profound influence on the generation and transfer of technology for small farmers. In Latin America, for instance, these linkages were accomplished by placing both R&E in a single organization. However, these linkages are still rudimentary and are not yet developed at all levels. A weak structure has led to a lack of adequate understanding of the diversity and complexity of the small-farm situation.

This is due, in part, to the fact that a relevant, closely linked R&E structure incorporating the Farming Systems approach to integrated rural development, in which priority is given to the labour-family-land relationship, is still lacking in most developing countries. It would seem that in most circumstances specific targeting on small farmer needs is absent or inadequate because farmer participation in research has not been treated as an important issue. The more the research-extension-farmer linkages are strengthened, the more opportunity it will give to research and extension services to understand small-farmer problems and to draw the attention of policy makers to the large potential for productivity that small farmers represent.

Three types of development are taking place in the field of linkage between research and extension. Functional improvements at the field level are being ensured through on-farm research initiatives. Structural improvements are being ensured in some countries in Latin America, either through placement of research and extension under a single institution or through committees with people drawn from research and extension, alternately. The third area of change is in functional improvements, where research and extension are considered as a continuum in the technology system and distinguished from programme or structural improvements. Effort is needed to ensure improvements in all aspects of the technology system, including policy and decision making, research generation, research transfer and research utilization.

Linkages can be effectively ensured only after four aims have been achieved: common interests and goals, mutual respect (e.g., between research and extension), mutual inter-dependence, and common funding. Moreover, joint planning can be satisfactorily ensured only through integrated funding. Effective linkages are therefore contingent on the condition that the motivation and goals of all those involved - politicians, researchers, extensionists and farmers - are consistent with each other. The goal of research may be to produce advanced technology, while the goal of small farmers may be to increase productivity without having to intensify the use of inputs over and above their means. Linkages can establish meeting grounds where the goals can be reconciled with one another.

R&E linkages can be strengthened through two channels. First all, extensionists involved in on-farm research can assist in on-farm trials managed by research and ultimately by the farmers themselves. (Such decentralization could also ensure multiplication in the number of trials). Secondly, research scientists could be involved as subject-matter specialists in support of the extension programme and adequately compensated to ensure their active interest and motivation.

A system of mechanical or functional linkage between research and extension has been attempted in India and found to be effective. Subject-matter specialists from the extension service are brought to a regional research station on the first two working days of each month, for Research-Extension Workshops. Research scientists at the regional station are responsible for providing training and guidance to the extension staff so they can perform the specific jobs necessary for various crops in the month to follow. In this kind of functional, face-to-face dialogue, goals of the concerned agencies converge, and working linkages between research and extension are established. This should be replicated, as appropriate, in other developing countries.

Increased farmer participation in the technology system as well as direct farmer involvement in government policy formulation is important, both in maintaining the flow of resources to the technology system, and in influencing the priorities and programmes of the technology system.

This would also help in ensuring the availability of adequate resources for small-farmer-oriented activities. Representation by outstanding and successful small farmers, who are often given official recognition by way of citations and prizes, may be an effective way of accomplishing this. Cooperatives are intended to serve as farmers' organizations, and it is also appropriate that these should be represented on research governing bodies at national, provincial and local levels.

Field days at the research stations, where farmers in large groups are brought together to give their candid reaction to research, can be effective in providing guidance to research. In this way, a relevant research agenda can be established with the help of the farmers themselves and not imposed from "above".

The R&E-farmer linkage should be two-way, ensuring delivery of technology from research stations to the farmers and feedback of farmers' problems to research, to make it more relevant to farmer needs. Participation of practicing small farmers may be most feasible at local research stations which are often linked to national research centres and, through them, to international research networks.

INVESTMENTS IN AGRICULTURAL RESEARCH AND TRAINING

The challenges facing agricultural research to develop a new technology which will be appropriate to small farmers throughout the developing world, are quite formidable. Generation of quality technology and a mechanism for its effective transfer are of prime importance to agricultural development. Rapidly increasing populations demand even more from a declining resource base. Risk-fearing farmers must be persuaded to introduce innovations into a traditional system. All this must be achieved against a background of fluctuating government policies and increasing budgetary stringency.

Research related to low-input technology suited to small farmers can be highly complex and sophisticated. The generation and adoption of these technologies need to take into account a complex interaction of social, ecological and economic aspects. This can be a very demanding process requiring considerable resources in terms of talent, time and finance devoted to basic research of a high quality in order to produce any results.

Investments in agricultural research, as a rule, have tended to fall short of the needs. Positive and stable government policy towards the agricultural sector in general, and towards the technology system in particular, is of critical importance for ensuring increased agricultural productivity on a continuing basis. The policy indicators are the level of government's overall financial commitment to agricultural research and to extension in particular.

In the context of the small farmer, a number of suggestions have to be considered. Sufficient resources need to be assigned to develop low-cost, low-external-input technology generation and transfer programmes. These should address the broad frame of social and economic change, including markets, prices, inputs and credit. Institutional and operational changes are required to ensure the integration of research and extension. In this context, on-farm research programmes offer an attractive possiblity.

Substantial investments in research are required to develop relevant technological innovations which can respond to the needs and constraints of the small farmer production system. This is becoming even more important in the context of the intensification of resource use, as demographic pressures make per capita land resources scarce, limiting extensive expansion of cultivable areas. In addition, environmental considerations are becoming a priority, and a great deal of agricultural research will have to focus on the development of a technology which is environmentally sound and ecologically safe. This would increase the pay-off from investments, in the long run. Moreover, the investments would result in a much wider set of technologically feasible options, all of which would not only enhance small farm productivity, but also protect its resource base.

In this context, biotechnology offers new options for the small-scale sector. Steps need to be taken to assure that the results from biotechnological research remain available for small farmers and the less-developed countries. To this end, much will depend on the adequate flow of resources to the relevant research institutions for the development of the necessary infrastructure to support biotechnology, both in industry (agro-industry) and in agriculture. Some measure of economy can be derived through linkages between international, regional and national research, and their coordination with private research. Because of the large investments involved, research in biotechnology and genetic engineering may tend to focus on large farmers. Care should be taken, therefore, to ensure that such frontier technology also serves small poor farmers.

The role of the private sector in generating technology (and sometimes assisting in its diffusion) should not be underestimated. Coordination of research activities in the public and private sectors is important in this regard, and steps should be taken at policy level to ensure such coordination. It is generally recognised that private sector involvement would greatly enhance the efficiency of the technology system. Managerial aspects are not the only important factors, however. For instance, the private sector may not always be equipped (or indeed have the incentive) to specifically focus on the specialised needs of various farming systems. National technology policy should provide incentives and subsidies to ensure private-sector research involvement in small-farmer problems and adequate institutional linkages would need to be established in this regard.

A long-term continuum of support for research which addresses a whole spectrum of small farmers' technical needs, is essential. Basic research and the biological sciences may provide the answers to the real problems of tropical agriculture (such as the development of crop varieties with tolerance to drought and salinity, the control of weeds and the discovery of new forms of resistance to pests and diseases) which have crippled small farmer productivity in the past. Some of this technology is already available, but commercialization has put them beyond the reach of small farmers. Hopefully, large scale investments, coupled with a favourable price policy, will drive prices down to levels which are within the means of the small farmer.

Similarly, there are some broader relevant technology-investment issues that have to do with the economic framework towards which a technology programme should be directed. Developing countries in Latin America, Africa, and in some parts of Asia, should give greater attention to agriculture than other countries in similar positions had done during the last few decades. Emphasis on investment in agricultural technology will also be dictated by a changing international economic setting. The industrialized world is likely to progressively decrease its trade protectionism and subsidy policies for agriculture, thus offering an opportunity for agricultural exports from developing countries. Therefore, today's low-productivity agriculture, which includes to a large extent the small-farm sector, has to be transformed (through appropriate economic policies) into being highly productive and geared both towards higher domestic consumption and exports.

If agricultural exports to developed countries were to materialize as suggested, research would have to face two divergent demands. On the one hand, there would be a greater need for problem-solving research to serve the subsistence requirements of the small farmers. On the other hand, there would be the demand for research to be sufficiently sophisticated to meet the conditions of international competition. This again emphasizes the need for increased investments in technology generation as well as its transfer to small farmers.

It is more likely that the focus will continue to be on food security rather than export, particularly for the small-farm sector. However, in either case, increased support for relevant agricultural research would be highly warranted.

In addition, research programmes need to be decentralized, and financial support provided, especially for operational purposes.

Investments are needed for all programmes, including research, extension, education and training. It must be noted that from within the limited resources available to programmes, research seemed to get more than its share of funding and the rest suffered in comparison. Moreover, within research programmes, social science research is not given adequate attention.

An important issue that requires attention is the need for valid data, particularly from small-farmer operations, so that strategies for dealing with their problems can be worked out on an informed basis. Therefore, funds also need to be provided for economic research and continued data collection.

Investment in Training is an important part of the technology transfer programme. Training is necessary at all levels for extension and delivery. This should include management training and the training of field agents in how they may effectively deal with farmers. On the other hand, farmer training can enhance efficient group formation, use of credit, development of small-scale enterprises, etc.

Similarly, the inter-relationship of public and private agencies(including academic institutions) has received little attention as a rule. How to improve the technical quality of research and extension is crucial, but a more fundamental need is the improvement of graduate-level training in developing countries. Research and extension services will be no better than the universities where people are trained to perform these activities.

Farmer organizations, group institutions and private agencies could help in relieving the strain normally faced by national governments with regard to funding research and extension programmes, but a strong policy has to be created to ensure their purposeful involvement in the small farmer context. There is skepticism, however, because some farmer organizations, such as cooperatives, tend to become politicized and dominated by large farmholders.

RELEVANCE OF AGRICULTURAL RESEARCH FOR SMALL FARMERS

Technology and the Small Farmer

Given the above characteristics of the small farm sector, the role of research is to generate technologies which can increase labour productivity and small farmer income. Agricultural technology is available from a number of sources - international research (including International Agricultural Research Centres of CGIAR), national research, scientific literature, private sector agencies and from the farmers themselves. None of these may have the solution to all the technological problems which small farmers face. However, an earnest effort is needed to focus on the small farmer context if the technology generated is to be validated and ultimately accepted by small farmers. The challenge is to address the constraints faced by the small farmers within their own environment, taking into account their limited resources and the fragility of the production system in which they often operate.

Given the precarious balance on which their production system often rests, small farmers are reluctant to take risks or accept additional investments. As a result, they have a natural, sometimes logical, resistance to change. In the light of their predicament, any given technology addressing small farmers must necessarily be based on a number of factors and variables, not the least of which is their limited resources. The development of new technology must also take into account traditional technologies used by the local small farming systems - a source which is often ignored.

The technology generation experience of research systems suggests that most of the agricultural technology packages developed are designed to suit large farmers and their resource endowments. The bias of research has been in favour of cash crops and its preoccupation with increasing productivity through its quest for high yielding production systems and the use of costly, high-input technology involving fertilzers with a focus on high potential areas, such as irrigated areas. Thus, the needs of small, resource-poor farmers in the marginal, rainfed areas have been generally neglected. Often, what is sought is a uniform blueprint technology which can be transferred to all potential beneficiaries, without resource-specific considerations. A single research package has not always been found adequate for meeting the needs of small farmers even though some research may be scale neutral. Farmers with resource-poor farms may not necessarily be poor: they may have off-farm income and their technology needs may be different from those of resource-poor farmers.

Technology generation with purely production oriented objectives needs to be reconciled with growth objectives based on principles of social justice and equity. There is a tendency for a downstream flow from research to extension to farmers but there are no effective mechanisms for the farmers to express themselves or to demand support. A balanced approach is required, taking into account both large as well as small farmers and their widely different technological needs. Research organised in this way would require increased small farmer participation in the technology system.

It is assumed in research circles that small farmers wish to remain confined to small scale activities related either to arable agriculture or livestock, but this gives research a narrow focus. The concern of small farmers for generating higher incomes and increasing their welfare must be fully recognised. Viewed in this context, it is understood that all their problems cannot be solved by agriculture alone and a much broader agenda is necessary.

Research should be focused not only on the optimization of production technology for the individual farmer but also on the technology needs of the entire household. Therefore, in solving the technology needs of the plant or animal in the small farmer context, his household needs (survival, food security etc.), must also be taken into account.

Another issue is whether research should be commodity oriented or generally oriented. Where rice yields have improved in Asia, rice prices have fallen, thus defeating the efforts to improve incomes through higher productivity. Research and extension need to help farmers in diversification - by creating new crops or livestock, a better use of water, etc. However, research and extension projects organized on a commodity basis have a limited capacity for addressing such issues.

Sometimes the above considerations have been taken into account through adaptive research, with direct involvement of farmers in the research process. Such research is known as cropping systems research, farming systems research or on-farm, client-oriented research (OFCOR). Farmer participation in OFCOR research is an important innovation, although adequate mechanisms still have to be evolved for sustaining it. Effective modes of collaboration and planning have to be developed between OFCOR and extension, while flexible strategies are needed for ensuring OFCOR's success.

An OFCOR programme (pioneered by the International Service for National Agricultural Research (ISNAR)) seeks to give the farmer a voice. It follows a problem-solving approach oriented towards resource-poor farmers, has an interdisciplinary perspective, and focuses on key problems facing small farmers, which have been identified through surveys and experiments in farmers' fields. OFCOR can ensure a direct link between researchers and farmers through farmers' direct participation in on-farm research, either on contract, as consultants, or through collaborative arrangements.

This approach has faced a variety of implementation and institutionalization problems, as the current OFCOR experience suggests. It requires interdisciplinary teams covering a wide range of biological and social disciplines. These are not easy to organize, manage or sustain, particularly because OFCOR involves research away from the research stations. In most cases, field research has been entrusted to generalists, supported by specialist farming system teams from research stations or from national programmes. Even where such teams are formed, however, obtaining cooperation from researchers is difficult because on-farm research is not normally part of the established research thinking.

Another problem faced in OFCOR implementation is the integration of OFCOR with research on experiment stations. OFCOR should complement the work on the experiment station and cannot survive as a stable component of the research system if it operates in isolation. Researchers are often critical of OFCOR because on-farm research is not done under controlled conditions. Moreover, conflicts in objectives have arisen when research normally deals with single crops as apposed to the full farming systems approach adopted under OFCOR. It is essential that OFCOR be attached to the research station with a multidisciplinary team within the station to guide the programme so that OFCOR activities are fully integrated with mainstream research. Another problem arises in the diffusion of the findings of on-farm research to farmers in large numbers. The OFCOR trials have a demonstration effect and some diffusion occurs through field days, but the impact is limited. Again, an effective linkage between extension and OFCOR is required through its involvement in on-farm research and with in-field extension to ensure the spreading of the on-farm research findings.

Many national agricultural research systems have initiated OFCOR activities but have had problems with institutionalizing the approach. Most of these are isolated programmes with substantial external financing. Integration of OFCOR with the established research system will require innovative mechanisms, but adequate national resources and policy support are neccesary for a sustainable impact.

In the light of the constraints faced by OFCOR, it was recommended that an OFCOR programme be backed by intensive training, which has implications for graduate-level training programmes in the countries concerned. Involvement of senior scientists, including social scientists and the integration of on-farm research with research work on experiment stations is crucial. OFCOR requires the strong support of senior management, sustained scientific leadership and the continuous training of the field staff.

It must be recognised, however, that research in developing countries is already being stretched from basic to applied research and to site-specific research, for which regional research stations are being set up in large numbers. In addition, farmer specific or on-farm research may stretch national research resources too far. In such cases, it may be more feasible to work through extension agents who are available in all areas and do simple on-farm trials. Farmer interaction with researchers could be ensured through field days and meetings with farmers at regional research stations or via feedback through extension workers. Socio-economic refinement of the research evolving at regional research stations could be ensured through follow-up on-farm operational research. Only fully tested results should be extended to farmers' fields. There is a growing concern that farmers should not be subjected to the risks of research. Detailed studies covering nutrition, sanitation, etc. are the means by which risks can be monitored. However, they may overload an already over-burdened national research system.

A research system directed at small farmers needs to resolve these issues in order to be relevant and effective. This can only be accomplished within a framework of a workable R&E link, which requires a strong extension input.

EXTENSION SERVICES

Investments in Extension

The rationale for ensuring adequate investments in agricultural research for the generation of new technology also has a bearing on the closely linked need to ensure its effective transfer. The challenge facing R&E in many developing countries is an enormous one, and technology systems, which are still in their infancy, cannot establish credibility unless a strong financial commitment to agricultural extension is accorded them.

Investments in extension tend to be restricted when technology is considered inadequate. On the other hand, pressures from extension should, in fact, help in generating research toward adequately improving the technology. Research and extension are parts of a single technology system and should be built up simultaneously. Experience shows that worthwhile technology can always be found for an earnest extension service.

A crucial issue which needs to be addressed in the context of extension investments is that of the direct (and indirect) benefits of extension. Extension systems have not been subjected to the sort of objective evaluation undergone by research systems. As a result, in ascertaining its effectiveness, research has been associated with high pay-offs, while extension has not been given the attention it deserves.

One tends to be oblivious to the many subjective advantages of investing in efficient extension systems. Increased investment in extension can bring about increases in agricultural efficiency. In this case, the society, and the economy at large, are the main beneficiaries. The consumer benefits through lower commodity prices and the trade sector benefits through increased agricultural exports. Lastly, the farmer benefits through increased productivity, efficiency and lower production costs. Thus, there is a valid case for increased investment in agricultural extension because agricultural extension is the engine of rural development.

In spite of this, in most situations an extension service does not seem to command a high reputation in government systems and unsatisfactory personnel get assigned to it. An extension service can be taken seriously by farmers only if it demonstrates that it effectively discharges its primary function - service to the farmers who are its clients. It will then begin to command satisfactory recognition within government, attract the required resources, and break the vicious circle.

Agricultural extension investments can be considered under four issues: (a) policy framework; (b) the extension infrastructure; (c) technological base of extension; and (d) sustainability. These are briefly discussed below:

(a) Policy framework: A national policy framework indicates production targets and priorities but rarely indicates the role of extension in the agricultural development strategy. This seems to arise either from ignorance or from confusion in the minds of policy-makers with regard to the role of extension in agricultural production. Clarity of objectives is unfortunately lacking within the extension service itself. This seems to be due to the fact that extension is frequently considered to be a backwater in agricultural service and often only unsatisfactory staff are assigned to it. The key issue is whether an extension service is capable of satisfactorily serving farmers and if it is not, it can only be considered irrelevant. This naturally leads to limited allocation of budgetary resources for the service.

(b) Extension infrastructure: The above attitude is reflected in the poor infrastructure within which extension has to operate. The main issue here is how to direct investments into strengthening the relevant components of extension infrastructure. Ideally, an extension organization and management should be simple, handling all production programmes through a single service, with short and direct reporting and supervision at the point of activity. Staff at all levels must clearly understand their duties, tasks and responsibilities. Staffing should be rational and appropriate to the task (i.e. the number of beneficiaries to be addressed and the extension message to be transmitted). It should also be consistent with the available communication media support. Reasonable salaries and incentives would encourage good work in the field. Adequate funds are needed to cover essential operational costs, and the employment of extension staff and their salaries should be complemented with adequate operational facilities for effective work and field orientation.

(c) Technological base of extension: The issue here concerns a relevant R&E linkage in which research must play its role in developing a solid technological base. This important input is frequently handled poorly. Extension must clearly recognize its dual function: to guide farmers, on the one hand, and to support the farmer-input service agencies on the other. Effective operational linkages between research and extension need to be established through joint planning, joint field visits, research-extension workshops, etc.

(d) Sustainability: Sustainability should be considered not only from technological and financial points of view but also from the participatory angle. A key element of sustainability is the orientation of the system to effectively ensure the participation of the intended beneficiaries. It must be recognised that the small farmholder is the client whose constant 'demand' for relevant technology/ message is the sole rationale for extension. Extension systems need to institutionally adapt themselves in appropriate ways to promote interaction with farmers and their active participation. The importance of farmer initiative in highlighting their problems and requirements is crucial. In the absence of such an intervention, the role of the extension system would be undermined, and consequently the justification for extension investments on a sustained basis would no longer exist.

Policy Environment

Economic policy which is frequently borrowed from neo-classical 'free market-oriented' models and adopted (inappropriately) in most developing countries tends to discriminate against agriculture by shifting the domestic terms of trade against it. Such policies have a detrimental impact on small farmers in particular. An extension service should have a role in influencing government policy to obtain economic policies that will make new production technology profitable to the small farmer. Social payoff to investments in extension may be low unless greater attention is given to the economic policy and principles which support the small farm sector and complement extension efforts. Sectoral policy discrimination against the agricultural sector in many countries also finds expression in shifting resources from agriculture to industry by lowering agriculture's profitability. Price plays an important role in raising agricultural production, but non-price measures (including credit, inputs, water and, more specifically, extension support activities) are also important incentives to increase productivity and agricultural output.

The Design and Functions of Extension Arrangements

National development policy frequently includes focus on increase in agricultural exports, growth with equity, poverty alleviation or rapid economic growth. These can result in conflicting demands on extension. While the clients for outward delivery systems are usually farmers, research, input suppliers, credit agencies and data analyzers are the clients of the feedback or inward delivery system. Information on farmers as clients and on technology availability are important supply issues in determining whether a campaign or participatory approach to extension is the most suitable one. Conflicts faced by extension as a result of national goals can be resolved through a balanced mingling of public and private involvement, in which entrepreneurs may take up the commercial opportunities, while public extension retains the major equity goal.

Four dimensions need to be considered when designing (or modifying) extension systems: goal orientation, time orientation, inter-personal orientation and structural orientation. These are directly related to the degree of certainty associated with clients' knowledge of their own requirements and the availability of appropriate technology to match these requirements. An authoritative, campaign approach may be appropriate if both are known, while a consultative or participatory approach may be required when one or the other of these parameters is unknown. If neither the client's perception nor the technology option is adequately known (as could be the case with rainfed technology), both the research and extension services and the farmers themselves would need to participate and interact in the "search" for technical solutions.

More and more educated people are being found in the rural areas as a result of improvements in educational facilities. This has important implications for the type of extension demanded. The higher the level of education in rural communities, the greater the number of farmers able to interact with the knowledge system. The education level of the farmers also carries implications for media communication support to extension. On the production side, there will be a demand for more sophisticated technology, even in marginal areas, which will have implications for increased extension costs. Extension systems geared towards biological control and integrated pest management are examples of areas where cost will be increased. The cost problem can be effectively dealt with by entrusting more functions to farmers and to farmer groups or by encouraging the private sector to seek farmer clients directly in promotion of recommended technology. As the scale of operations increases, the unit costs will decline.

In a similar context, extension for small farmers who number in the thousands, may, in most cases, be made more effective if farmers' organizations can be formed for the purpose of facilitating broader interaction with extension agents and researchers. Agricultural cooperatives should be encouraged so that extension agents may deal with farmers in groups rather than as individuals. The cooperatives can take care of credit delivery and recovery, inputs, distribution and marketing functions.

There are three definitions of extension activities: agricultural production services (or income generating activities); rural community development, and comprehensive non-formal continuing education. These definitions reflect organizational choices based on the functions to be performed. In the Third World, most extension systems adhere to agricultural production services.

The major function of extension is to enhance or facilitate the communication of practical knowledge on agriculture or rural development, usually as "non-formal" education. Extension has to deliver technology, facilitate educational delivery and problem solving, ensure skills development, encourage feedback and be involved in adaptive research or farming systems research. All of these have implications for the training, management and organization of extension.

Improvements in extension should be directed towards a number of specific programmes. Firstly, there is need for agricultural extension management training at all levels. It is important for policy makers to become more aware of farmers' problems. Secondly, all available local resources - including the private (informal) sector - should be mobilized for developing extension. Thirdly, farmers involvement and influence on extension systems is a paramount consideration. This will make extension more relevant to farmer needs. Fourthly, there is need for supporting research and data collection on extension, which is very rare at present.

As mentioned earlier, it is essential that farmers have a role in defining agricultural policy and programme priorities. Extension has an important role in building this up, with the long-term goal of giving an effective voice to the producers of food. (Farmers' advisory committees have a great deal of influence in the USA. There is no reason why this cannot be encouraged in the developing world also).

Recent developments indicate that a new environment is emerging in the context of extension systems and various models for the transfer of knowledge have been developed to address a number of issues. These include: i) the focus on the private-sector; ii) the privatization of certain public extension systems; iii) the trend among large farmers to by-pass public extension services; iv) the effort by certain research institutions to provide what has been referred to as frontline extension; v) the development of new designs and mechanisms for linking research and extension, vi) the search for participatory methods; and vii) the experimentation with hybrid research-extension systems.

There are different types of national arrangements within the various extension systems. Public extension may be organized directly, or through parastatals, or even jointly with farmers' associations (as in Taiwan). Private extension may be organized for profit (e.g. cooperatives or large estates), by multinationals, non-profit organizations or by membership organizations such as farmers' associations. The mix of arrangements most valuable for any country will depend on the policy goals of that country. A careful multi-disciplinary diagnostic study would determine the correct mix of public, private and mixed services.

There is also scope for activities such as front- line extension (e.g., through agricultural science service centres in India, which play a significant role in the vocational training of farmers). National demonstrations by researchers are an important research-extension function. The demonstrations ensure a direct feedback of farmer problems to research. Flexibility is essential in extension programmes, especially those which suit specific situations. (This places a demand both on the national governments and on the external funding institutions to identify the specific needs in each situation.

Sound research is essential for sustaining extension: unless there is worthwhile technology to offer, extension will not succeed. Research stations can provide considerable support to extension work through trials and demonstrations set up by the research stations. Local research sub-stations can be useful in directly solving farmer problems, so farmers should be encouraged to visit these sub-stations.

If a system such as T&V focuses on the delivery side of its function alone, it operates as a campaign approach. As discussed above, this is successful only in cases where both the relevant technology and the client's needs are known. When one of these is not known, a campaign approach will fail. This emphasizes the essential feature of the feedback function of T&V which is critical for its success. The T&V system can be considered satisfactorily organized <u>only</u> when both the delivery system and the feedback system have been effectively developed and implemented.

A campaign approach will work only in cases where the client is known and the technology is available: the T&V system is most functional in such a situation. But in the case of small farmers growing rainfed crops in marginal conditions, neither client nor technology are adequately known. Would the T&V structure be adequate under such circumstances? By most accounts the answer would appear to be negative, provided that the T&V system is augmented by a participatory approach which takes into account a measure of "client demand".

T&V is presented as a system which is adequately equipped for dealing with various extension situations because it is supposed to be a composite of delivery system and feedback. In reality, however, it often happens that the 'feedback' element in T&V is inadequate, in spite of the frequency of the 'visits' because of the emphasis on a one-way flow of information through 'Training'. A longer time-horizon is needed for understanding the rainfed client and identifying the best technology suitable for him. The T&V system also needs to pursue a more participatory approach, for which the system is inherently equipped.

The time horizon of the resource-poor farmers is often limited to the crop season. On the other hand, the time horizon of the T&V extension agent is even further restricted to the fortnightly training and visit cycle. The extension agency must retain a focus on the seasonal requirements of crops without which the principles of T&V easily breakdown.

The T&V system is based on the group teaching approach and does not give adequate weight to mass communication methods such as radio, television, leaflets, etc., which would improve its (cost) effectiveness, in many situations, especially where the media is very popular and there is a fair measure of literacy.

In many countries it appears as if the role of extension in general and T&V in particular is not adequately accepted. What is being questioned is how effectively the extension service is able to discharge its role. In Africa, for instance, one of the issues is how to adapt T&V to local conditions. A number of adaptations to the T&V system are being attempted: use of groups, communication support, farm trials, research inputs, etc. The basic principles of T&V are important and should be sustained. However, appropriately trained and motivated extension staff fall short, at present, of what the system demands in order to address its tasks effectively. The T&V system serves as a vehicle for technology to farmers and should also serve as a vehicle for feedback of farmer problems to research. However, delivery of technology must also result in concrete gains through adoption of the given recommendations, in order that a feedback cycle can be encouraged and the T&V system made more meaningful and effective.

EXTENSION SUPPORT

The above overview of extension systems raises issues of financial and institutional viability. Extension cannot be effective if divorced from its economic viability. There is a need, therefore, to strengthen the economic core of extension programmes, including marketing, production economics and information.

Off-Farm Activities

There is a tendency to view farms as static receptacles for new production technology. That, however, is not how the system works, in many instances. New technologies may have a differential effect in enhancing resource productivity. Some technologies may raise productivity of land, others of labour. Increased per capita incomes in the non-farm sector, resulting from general economic development, may result in an exodus of labour from agriculture. This may also happen when demographic trends exert pressure on farms and increase the demand for non-farm employment. Off-farm activities in rural areas can offset rural-urban migration. Although national governments should be concerned about this, there is also a role for the extension service in this regard.

Transfer of labour out of agriculture could result in a costly geographical move, both socially and economically, if the labour has to be absorbed elsewhere. A more rational policy would be to generate the new employment opportunities in the vicinity of the surplus labour. This implies that extension should give high priority to rural community development programmes in marginal areas with limited agricultural development opportunities on the farm.

However, there are a number of issues that extension must resolve in the context of farmers on marginal or inherently unproductive lands, or in situations where the cost of ensuring sustainable agriculture may be high. In such situations, efforts may best be directed to inducing the population off the marginal lands and into alternative employment. However, marginal lands are often found in economically deprived regions where creation of off-farm employment is difficult. In the interest of inter-regional equity and rural welfare, infrastructural investments in marginal lands may be desirable in such situations which can lead to the promotion of appropriate economic activities. The role of extension is essential in the promotion of these activities, which have a traditional advantage.

Considering the magnitude of the problem of surplus manpower, migration investments in programmes for off-farm employment within the region will be less costly than a large-scale resettlement programme. The suggestion that out-migration from marginal agriculture should be contained within the area is, however, a challenging proposition, calling for development of infrastructure, access to raw materials, and the capacity to compete with more developed areas. On the other hand, when handled adequately and with the appropriate support of extension services, such areas have in the past been transformed into economically potent zones. Extension support activities need to be organized to address these broader purposes.

Skills development and education are more important than the mere delivery of messages, and extension support activities should be equipped to ensure this. The focus should be on the development of the individual through vocation-oriented education. There is a tendency to underrate the capacity and potential skills of individual farmers in an underdeveloped setting. There is often, however, plenty of scope for skill development among such farmers, and this needs to be exploited for the farmers' own wellbeing as well as that of their production system.

In this context, the T&V extension system has been seen as an effective means of revitalizing a moribund extension service, but it may be too labour-intensive and costly in the context. However, T&V is confined only to the delivery of technology messages and has little focus on skills development. Extension must be educative, i.e. it must teach principles and assist small farmers through their production cycle and beyond and should not be merely a means of providing information.

The question of focusing research on marginal lands and sustainability is a crucial issue. The social and economic cost of not pursuing economic activities on marginal lands has to be taken into account when considering investments in extension support in marginal lands. This is important in the context of environmental protection and sustainability. Many such areas have high population pressure, and absorption of surplus labour within the region could be a very long term exercise requiring the development of new technology for this crucial factor of sustainability. Meanwhile, action needs to be taken, at least in order to arrest the degradation caused by intensive land use. The tendency to deal with these issues merely as a technology problem within agriculture has a low social payoff; an approach beyond agriculture would be more effective.

TECHNOLOGY UTILIZATION FOR SUSTAINABLE AGRICULTURAL DEVELOPMENT

In a number of small countries, especially the less developed ones, considerable scope exists for improving the transfer of comprehensive, viable technological packages (including the provision of institutional support) for agricultural development. Alleviation of poverty in the rural economy of such countries and regions requires that:

- the new technologies on crops and production systems introduced in most environments should focus on increased labour productivity and expanded employment;
- new crops and technologies take into account the complex patterns of seasonality in production and processing, both of which impinge upon the opportunity cost of labour, and require investment in locally applied agricultural research capacities;
- the introduction of new technologies will not require large amounts of working capital on the part of the farmer until viable and suitable credit and other rural financial systems are set up, because liquidity constraints and high-time preference rates of the poor farmers in the subsistence economy will prevent technology utilization. (This is especially true for women farmers who lack access to credit to finance technological change);
- the marketing channels for inputs and outputs of newly-introduced crops must not be excessively risk-prone (There is a key role of credit and savings assistance in coping with risks);
- new technologies and crops take into account the agricultural resource base and be environmentally sound, i.e. they should ensure that soil fertility and sustainability of agriculture will not be at risk by their introduction;
- production, consumption and health environments be inseparably tied together in the rural economy and have to move together. (Rural infrastructure and services play a critical role in interaction with technology utilization for sustainable agricultural development).

Thus new technologies can be utilised more efficiently if a number of policy, institutional and technological issues are resolved and sustainable agricultural development can be pursued.

POLICY AND INSTITUTIONAL ENVIRONMENT

Public policy has to assume increased responsibility in creating an environment which is conducive to technology utilization. Only then can well-designed research systems be effective in generating new and appropriate technologies and extension systems be effective in ensuring their transfer to farmers.

Appropriate broad-based policy and technology generation through programmes for the spread and utilization of technology, complement each other. However, one must not lose track of the specifics required to make technical change work in favour of the poor. Critical issues in this regard are the explicit concern for women farmers' labour productivity, their access to credit, their drudgery in performing a variety of activities, the risks associated with pricing and production under new technology utilization, the sustainability of production under the changed technology and the constraints of small farmers in financing utilization of that technology. The latter requires that increased attention be given to viable rural financial institutions catering for the poor. Of significant importance are the programmes and projects that support the build-up of local and national capacities in institutions which can effectively tackle these issues.

As the process of technology generation and transfer becomes increasingly complex, both at institutional and farm level, improved education becomes increasingly relevant because it enhances the potential for the success of technology utilization. Investment in rural infrastructure provides the basis for the availability of goods, services, and production inputs required for accelerated technology utilization. It also facilitates diversification toward new crops (food and non-food) and the gains from regional specialization and interregional trade. As mentioned earlier, household food security is a major consideration in Africa and Latin America, but nutritional adequacy concerning poor small farmers has become a major concern in Asia. Diversification of agriculture has, in some cases, had an adverse effect on nutrition, e.g. through neglect of traditional crops like coarse grains, roots, tubers and pulses. Vertical diversification may, on the other hand, be more productive than replacement of traditional crops, e.g. through value-added production of crops such as palm oil, coconut and rubber - some of which concern small farmers. Extension should be concerned about the importance of the profitability of the recommended technical packages they extend.

National plans generally include platitudes about concern for marginal areas in policy statements but often do not provide any concrete programmes to deal with the problem, simply because the institutional and technological issues are so formidable and they require bold policy intervention, something which is rarely exemplified.

The Marginal Lands

Population growth, along with the need for increased output in a situation where there is relatively limited availability of arable land, establishes the need for innovative technology generation and utilization in agriculture, in order to achieve sustained growth in agricultural production. Higher food production, too, generally means intensification of agricultural production and accompanying resource degradation. In order to be of lasting benefit to the poorest segments of the rural population, new production technology must raise labour productivity in ways which are environmentally sound. It must also increase employment, and thereby effectively raise the demand for food. In order to effectively alleviate poverty, new technology must address those rural households which have limited resources and do not produce a marketable surplus, as well as the landless population who are both capable and willing to provide farm labour, make use of appropriate technology and generate income.

The marginal or neglected areas have a unique mix of poverty and environmental problems requiring special attention. Research is weak in developing sustainable technologies for these areas, thus further complicating the development effort. The suggestion that these areas may be provided with inputs free of charge or on the basis of subsidies does not appear to be a sustainable approach. Experience shows that such programmes will have a short life-span, often that of the external funding for the programme. Areas with high production potential which have gained from extension services should be made to pay for it, and funds derived thus may be channeled to support extension work in low-potential areas. M&E is useful in the context of evaluating the possibilities of cost recovery for extension and determining the feasibility of this in resource-poor countries with wide regional disparities in resource endowments.

Cost effectiveness of an extension programme is another important issue. It is easy to assess the costs of an extension programme but it is usually more difficult to envisage the benefits it will bring. It is, therefore, a frequent misconception that extension for poor small farmers is all cost and no benefit. Whereas there may be limited (tangible) monetary benefits from small-farmer extension, social costs and benefits in terms of political stability and long-term ecological effects also need to be taken into account. Giving credence or support to extension activities ultimately becomes a matter of political decision where social profitability should be a major consideration, although this is often not the case.

Another key issue is whether the extension programme is being run effectively or not. Specifically, the adoption of recommendations made by research, through noticeable increases in farmer earnings, needs to be assessed. Once the extension programme is successful in ensuring these results, farmers themselves will speak in defence of extension. M&E should bring out such merit on the part of the extension services and highlight the positive elements of extension which are not readily perceived. Innovative techniques need to be developed where concepts are subjective in nature and quantitative assessment is being performed. The evaluation should further be able to ascertain whether the technology promoted was adaptable or not and should be capable of discerning the factors responsible for package uptake or rejection.

Attention to research and extension work for women farmers is emerging as a major concern, but little effective effort seems to have been made in this regard within the developing countries. This needs to be monitored with a view to enhancing R&E effort towards the specific problems of women farmers and their critical role in rural households as providers of adequate nutrition/food security and in supplementing family income.

Much has been written about the difficulties of identifying specific small farmer needs and the extension approach which may be most suited to them. The various systems should be evaluated so that necessary judgements can be made. In this regard, it has been widely recognized that no satisfactory means for monitoring and evaluation of extension projects has so far been set up and that this is a weakness across the board. IFAD is one of the institutions attempting a great deal in this regard but there are problems because of the paucity of appropriate methodologies and indicators. Gradual improvements are being attempted in project design, based on whatever information is generated through the implementation of those projects.

CONCLUDING REMARKS

The foregoing chapters of this book have highlighted the different critical aspects of the Technology System. The major conclusions which can be drawn are the following.

Firstly, it is increasingly recognized, not least due to IFAD's own efforts, that the technology needs and requirements of small farmers should not be overlooked and that improved technology needs to be developed and tested within their socio-economic conditions. Moreover, the role of small farmers themselves, in assisting the development of sustainable technologies, should be taken into cognisance. An appropriately interactive linkage with farmers' organizations can be most useful in bringing this about effectively. Such grassroot groupings can become not only a vital source of traditional technology but also reflect a vital pressure group in society, capable of articulating their needs and seeking a redirection of the focus of technology systems to respond to the requirements of small farmers.

The role of R&E in bringing frontier technology (for instance, in the field of biotechnology - molecular biology and gene manipulation), combined with improved traditional technology, to the door-step of the small farmer is crucial, though nonetheless challenging. There is a clear need to identify relevant modalities of R&E linkages. What is needed is a meaningful triangular interaction between research-extension subsystems and farmers so as to give a more dynamic focus and encourage greater collaboration between researchers and resource-poor farmers in technology generation.

Secondly, the issue of effective transfer systems and the Research and Extension linkage in particular, is of paramount importance and various institutional issues identified in this book have yet to be resolved. The focus needs to be placed on the role played by the 'Extension' system, not merely as an end in itself but as a means to an end. In this context, the economic benefits which may accrue from the system warrant adequate evaluation and need to be clearly brought out in order that it provides the evidence for governments to justify greater resource allocation to this crucial aspect of the Technology System. Monitoring and evaluation units within research and extension subsystems should be strengthened with internal management tools in order to assist in this. They should be able to provide useful analyses, objective assessments of extension performance, data and critical information to continually adapt R&E systems to the needs of its clients - the small farmer.

Thirdly, there are no infallible modus operandi for organizing an extension system. However, it is recognised that it needs to be dynamic and flexible. In this context, various approaches to Extension such as the T&V System and its variants need to be reassessed in relation to the situation-specific requirements, in order to create an efficient transfer system with a truly participative character. The extension system so identified, must be such that the basic attributes of a professional extension service, including the ability to transfer high quality technology provided by research and systematic farmer contact, are not compromised.

To achieve this will entail a highly resource-intensive Technology System, the cost of which is often prohibitive and beyond the means of developing countries' recurrent budgetary resources. The involvement of the non-public sector becomes crucial in this respect, and the maximum possible support of the private sector, commodity boards, cooperatives, NGOs, etc., needs to be harnessed for a more effective impact.

Fourthly, in recognition of the vital multifarious role undertaken by women in food and cash crop production, processing and marketing activities, livestock management and, most important of all, in improving family welfare and household nutritional status/food security, R&E needs to have a special focus on gender issues. Greater emphasis is required on designing time and labour saving technologies for women, in improving their access to credit through adequate arrangements for institutional financing, in introducing collective facilities to reduce their drudgery and in identifying alternative sources of water/firewood within easy reach of households.

The above set of considerations raise policy issues which must be addressed if the Technology Systems are to achieve the desirable objectives of sustainable development. For instance, new production technologies generated and transfered by R&E systems must not only concentrate on raising labour productivity but also on creating avenues for both on and off-farm employment, thereby effectively raising real incomes while reducing demographic pressure on limited farm resources and promoting the sustainability of the production system. For this to be realised, national policies need to recognise the importance of allocating greater resources to R&E subsystems in a way which is commensurate with the requirements for their sustainability. Necessary resources and institutional support for strengthening R&E will also facilitate meeting the goals of national self reliance and increased income and food production. A strong policy environment also needs to be created in a manner which is conducive for small farmers to adopt appropriate technologies.

At the same time, a balanced all-round development would entail parallel investments in rural infrastructure, including marketing facilities. This will enhance the potential for the successful adoption of improved technologies, through the availability of goods, services and production inputs for technology utilization. Policy measures are required to introduce instruments for the provision of adequate inducements and incentives for Technology Systems to cater for small farmers' requirements. In this context, R&E needs to be focused more strongly, not only on high potential areas, but also on highly risk-prone rainfed agriculture in marginal areas as well, since a substantial rural poor population subsists in the harsh conditions imposed by the latter.

Hopefully, this book will generate interest in resolving some of the central issues which impinge on Technology Systems for small farmers, on the part of researchers, administrators, and policy makers alike. There is a need to consolidate the positive achievements, eliminate the negative elements encountered in development interventions and use the experience which has emerged, to design better and more effective rural development projects involving the transfer of appropriate technology. The goal would be to achieve a process in which research and extension can develop suitable Technology Systems, with a pivotal role played by the farmers themselves, to ensure sustainable, integrated rural development.

ABOUT THE AUTHORS

Dr. Abbas M. Kesseba, a graduate of Cairo University, holds a Ph.D. degree from Munich University, West Germany, where he pioneered in research on the dynamics of enzymes in soil bio-chemistry. Dr. Kesseba was Professor of Agricultural Chemistry at the University of Dar-es-Salaam as well as Dean of Post Graduate studies at the Universities of Kenya and Zambia. He was Advisor to the Egyptian Minister of Agriculture and was Egypt's first Permanent Representative to the UN Food and Agricultural Organization (FAO).

Dr. Kesseba was the Coordinator of the International Fertiliser Supply Scheme (FAO), and was the Chairman of the FAO Agricultural Committee and the FAO Intergovernmental Grains Group. He played an active role in planning the UN Conference which established the International Fund for Agricultural Development (IFAD), where he is currently the Coordinator of the Technical Advisory Unit.

Dr. Kesseba has published extensively on the subject of biological research and specifically on soil science and agricultural chemistry in numerous international scientific journals. His publications include a book on fertilizer policies for the development of the agricultural sector in Egypt.

Mr. Shantanu Mathur holds two graduate degrees in Economics and Mathematics from Allahabad and Cambridge Universities and two Master's degrees from Delhi University and Cambridge University in Pure and Applied Economics.

Mr. Mathur has worked as an Economist at the UN Food and Agricultural Organization (FAO) and, between 1985/87, contributed to the development and application of several models including the FAO Agriculture: Toward 2000; CAPPA Model for Agricultural Planning and the Livestock Development Planning System.

Mr. Mathur has done field work for the International Fund for Agricultural Development (IFAD) in The Gambia, Lesotho and Botswana as Project Economist and is currently at the Technical Advisory Unit of the Fund providing research assistance in its operations related to project work and Technical Assistance Grants. He has published papers on the subject of development economics, and has been the author of several FAO and IFAD reports as well as contributed to the FAO Book, Dynamics of Rural Poverty.

Dr. Martin E. Pineiro has a degree in Agricultural Engineering from the University of Buenos Aires, a Master of Science in Agronomy from Iowa State University, and a Ph.D. in Agricultural Economics from the University of California at Davis.

Since 1986, Dr. Pineiro has been the Director General of the Inter-American Institute for Cooperation on Agriculture. Prior to that he was Research Coordinator for the Centro de Investigaciones Sociales sobre el Estado y la Administración (CISEA) and for the Service for National Agricultural Research. Dr. Pineiro has also served as Undersecretary in Argentina (Secretaría de Estudio de Agricultura y Ganadería).

Dr. Pineiro has published various articles and books related to technical and social changes in the economy and agriculture, technical innovation and agricultural research in Latin America and the Carribean.

Dr. Alexander von der Osten, a German national, is Executive Secretary of the Consultative Group on International Agricultural Research (CGIAR). Previously he was with the International Service for National Agricultural Research (ISNAR) as Director General. He has wide experience in international development, including having been in charge of research projects at the University of Heidelberg carried out in conjunction with FAO. He has served as Executive Secretary of CGIAR's Technical Advisory Committee and was a member of the FAO Director General's Program and Policy Advisory Board. Mr. von der Osten has held senior positions at the German Agency for Technical Cooperation and with ISNAR, prior to his appointment as Director General. In Germany he was Managing Director of Weingut Eduard Diehl, an agriculture-based enterprise in the private sector.

Dr. Peter T. Ewell, an Agricultural Economist, received his Ph.D from Cornell University. He has worked as a consultant at the International Service for National Agricultural Research (ISNAR) in the Netherlands on the core staff of a comparative research project on on-farm, client-oriented research in nine national agricultural research systems. He is currently the Regional Social Scientist for Africa at the International Potato Center (CIP).

Dr. Merrill-Sands, Research Officer at the International Service for National Agricultural Research (ISNAR), Netherlands, received her Ph.D. in Anthropology at Cornell University in 1984 with a dissertation on mixed subsistence-commercial farming systems in Yucatan, Mexico. After working as a consultant to the FAO in Rome and to the Technical Advisory Committee of the CGIAR, she joined ISNAR in 1985 as a research fellow under the Rockefeller Foundation Social Science Fellowship Program.

Since 1986, Dr. Merrill-Sands has been project leader of the nine-country study on the organization and management of on-farm client-oriented research in national agricultural systems.

Dr. William M. Rivera, Associate Professor, Department of Agricultural and Extension Education, University of Maryland, College Park (UMCP).

Dr. Rivera has many years of experience in Adult, Continuing and Extension Education and has worked internationally with UNESCO, OECD, World Center for International Extension Development (CIED). In 1985, he initiated at UMCP a colloquium series on "Agricultural Extension Worldwide" which has trends in agricultural extension. In addition, he is the main organizer of the Annual Lifelong Learning Research Conference. Active in several professional associations, he is co-chair of the U.S. International Science and Education Council's Committee on International Organization Affairs. In 1987 he received the Gamma Delta award for outstanding work in international agricultural extension.

Dr. Rivera has produced over 70 publications on adult and extension education, including: Designing Studies of Extension Program Results (1983, 2 vols.), Comparative Extension (1986), Planning Adult Learning (1987) and Agricultural Extension Worldwide (1987).

Mr. J.L. Bajaj joined the Indian Administrative Service (IAS) in 1962. As part of the Civil Service he has held many senior positions in the Indian State of Uttar Pradesh, including that of Joint Secretary and later Secretary, Ministries of Planning and Finance.

Mr. Bajaj was awarded a fellowship at the Centre for Development Economics, Williams College, USA, for Post-Graduate Studies in 1970-71. Mr. Bajaj has also been the Director of the Administrative Training Institute, Nainital, India, and the Resident Director of the Indian Investment Centre in New York.

Dr. John A. Hayward has a Masters degree in Entomology Research from the University of Nottingham and a Doctorate in Agricultural Biology from the University of Reading.

He has worked for fourteen years in West Africa doing research on crop protection and was the Senior Officer of the Cotton Research Corporation in Nigeria.

Dr. Hayward was a member of the Professional and Academic Board of Ahmadu Bello University in Nigeria and was heavily involved in developing farming systems research in agricultural extension and in training at all levels from university to farmers.

Since 1980, Dr. Hayward has worked for the World Bank in Washington, D.C., focusing on the technological and institutional aspects of agricultural development in Asia and Africa. He is now the Bank's Advisor on Rainfed Crops and is currently responsible for formulating Bank policy on agricultural extension.

Dr. Michael Baxter, before joining the World Bank in 1977, worked in the Pacific and Brazil for seven years, primarily on issues on farmer organization and land use. At the World Bank he has worked in agricultural development, particularly the Organization of Farmer Support Services, and is presently Head of the World Bank's Agricultural Unit in New Delhi, responsible for on-going Bank irrigation and agricultural projects in India.

Dr. Joachim von Braun graduated from the University of Bonn and received his Doctoral degree in Agricultural Economics from the University of Göttingen.

Dr. von Braun is currently the Project Director for Food Security at the International Food Policy Research Institute (IFPRI) in Washington, D.C., where he has been employed since 1984. Prior to that, Dr. von Braun was a Research Associate and then Assistant Professor at the University of Göttingen. He was also a Visiting Fellow at the Institute of National Planning, Cairo, Egypt.

Dr. von Braun has conducted numerous research projects in developing countries including Turkey, India, Bangladesh, Syria, Haiti, Rwanda, and Ethiopia. He has also published extensively on agricultural development and economics.

Dr. G. Edward Schuh is Dean of the Humphrey Institute of Public Affairs at the University of Minnesota. Prior to that he was Director of Agriculture and Rural Development at the World Bank, Washington, D.C.

Dr. Schuh has a B.S. from Purdue University, an M.S. from Michigan State University, and an M.A. and Ph.D. from the University of Chicago. He was Professor of Agricultural Economics at Purdue University (1959-79), and Head of the Department of Agricultural and Applied Economics, University of Minnesota (1979-84). He was also Assistant Director of Extension during part of the period.

Dr. Schuh has also been Program Adviser to the Ford Foundation in Brazil, senior staff Economist with President Ford's Council of Economic Advisers, and Deputy Under Secretary of Agriculture, U.S. Department of Agriculture. He is a past President of the American Agricultural Economics Association.